Weather Things you Always Wanted to Know

The inside story on the outside story.

Alan Sealls

*The Best Weatherman Ever**
**According to the internet!*

Alan Sealls

An Intellect Publishing Book

Copyright 2023 Alan Sealls

ISBN: 978-1-961485-17-4

First Edition: 2023
FV -10 HB

www.AlanSeallsAuthor.com

Intellect Publishing, LLC
6581 County Road 32, Suite 1195
Point Clear, AL 36564
www.IntellectPublishing.com

Introduction   5

  1.   *Sun, Air, Wind, and Atmosphere*   *11*

  2.   *Water Cycle and Hydrology*   *37*

  3.   *Clouds*   *67*

  4.   *Seasons and Weather Cycles*   *87*

  5.   *Radar, Satellite, and Instruments*   *107*

  6.   *Lightning and Thunderstorms*   *123*

  7.   *Tornadoes*   *149*

  8.   *Hurricanes*   *177*

  9.   *Sky Sights and Optics*   *223*

  10. *Climate and Climate Change*   *239*

  11. *Winter*   *273*

  12. *Weather Forecasting*   *293*

Final Thoughts   323

Alan Sealls

Introduction

Meteorology is simple, yet complex, with amazing details that only become apparent on close study. The beauty of a single snowflake yields precision-perfect geometry, yet no two snowflakes are identical. Similarly, it is tiny and sometimes imperceptible differences in weather parameters which make no two weather scenarios the same. Trace the path of a leaf in a stream and predict where it will end up. That's the challenge of a weather forecast and the nature of nature. Different ratios of heat from the sun, air, and moisture make all weather, often visually announced by clouds. Clouds, like people, come in different sizes, shapes, colors, and heights, but all are made of the same thing—water.

Water is the foundation of life. Air is required, too, for us humans. Look up in the air and use some imagination. You might envision molecules moving around in their daily hustle and bustle, hooking up, separating, creating life-supporting H_2O, or plant-sustaining CO_2. Of course, too much CO_2 is a long-term issue.

Seasons offer up modes of meteorology. You may take in a crisp, calm winter week followed a month later by a springtime dazzling, delightful day. Maybe you enjoy a serene, summerlicious afternoon before you feel the frosty fingers of fall. Weather is universal, dynamic, and ever-present. Weather principles are a part of our lives and language. Have you had a heated discussion about your thermostat? What's yours set at, 68, 72, 75? How many times did you debate, negotiate, and then agree to leave it at a specific temperature only to go back later and see that it mysteriously was changed? Maybe you need to use a camera to catch the culprit.

Is that a gale-force wind? Who cranks the ceiling fan to top speed, generating high pressure and blowing papers all over the room? Who fails to turn on the exhaust fan, which is designed to create low pressure and draw humidity and odors out of the kitchen, or bathroom, like a waterspout lifting sea spray?

Who demonstrates the water cycle, contributing condensation and a higher dew point temperature in the house? Whoever languishes the longest in a hot shower is guilty. The clouds of vapor left behind are evidence. However, that may not be the same person that lets the bread get soggy. That person forgets that if you put the bread next to something really cold, the moisture in the bread condenses.

Someone prefers every light in every room in the house on to give the impression of high noon in the

Sonoran Desert. If the bulbs are the older incandescent type, that's a lot of heat.

Maybe a bedmate sets up the global warming scenario every night, in piling on blankets, trapping heat. The other is forced to adapt to the changing climate by shedding garments. Maybe that was the plan?

Who opens the shades or blinds to let the sun in? That's good in winter but not so great in summer when the sun quickly heats your house, or if you are in your birthday suit.

Who spends more time holding the refrigerator door open, thinking, deciding, daydreaming? It doesn't cool the entire kitchen, but it does make the fridge work harder, and it raises the electric bill a bit.

Do sweaty relatives visit too often, hurling hot air in too-long conversations? That sounds like relative humidity.

Who knew there was so much meteorology in matrimony and homelife?!

My third-grade teacher, Miss Costello, took our class outside and we laid on our backs in the grass and looked up at the cumulus clouds to try to imagine the shapes that they reminded us of. We saw animals and dragons and people. That was called nephelococcygia by the Greek playwright, Aristophanes. Set the meteorologist inside of you free. Don't be embarrassed! You don't have to know the names of clouds to appreciate them.

How many lyricists have made meteorological observations? "I can see clearly now, the rain is gone." "Summer breeze, makes me feel fine." "I see skies of blue, and clouds of white." "Feel the rain on your skin. No one else can feel it for you." Rain can make rainbows. Surely, you get a good feeling when you see one. The natural world offers entertainment, inspiration, challenges and treats to the eyes and mind, but you have to open both of them. My hope is this book provides the inside story on the outside story, answering questions you've always had about weather.

Weather Things you Always Wanted to Know

The inside story on the outside story.

Alan Sealls

CHAPTER 1

Sun, Air, Wind and Atmosphere

Where can it rain but never gets any wetter?
The ocean.

Weather is the day-to-day changes in the atmosphere. From serene to extreme, from dreary to eerie, from sublime to bizarre, weather is the motion of meteorological elements. Not earth, wind and fire, weather is sun, water, and air. That's it. Those 3 simple ingredients cook up a buffet to serve the planet with variety.

Without the sun, no air would move on Earth. The sun is our nearest star, delivering barrels of buttery sunshine. It is the primary source of energy for our planet. Averaging 93 million miles away from us, the sun sends light on an 8-minute journey, along with heat, energy and other radiation through the vacuum of space. It is the engine for weather, causing water to change phase from ice to liquid to vapor, and forcing air to expand. The sun heats Earth unevenly. Direct rays nearer the equator keep that part of the globe warm, while a low sun angle at the poles doesn't provide as much heat. Aside from sun angle, in any given region of Earth, heat absorbed from the sun depends on the type of land or water surface, cloud cover, and snow cover. Heat from the sun makes air move as wind.

The rotation of Earth on a tilted axis around the sun gives different amounts and duration of heating throughout the year. This is the reason for seasons. In the summer, the sun is high in the sky, making the rays more direct and intense. From the time the rooster crows to the time the sun goes to bed, summer daylight hours increase. The winter sun is low and weaker, and the hours of daylight are fewer. In fact, at the poles, there is no sun in the winter, but there is 24 hours of sun in the summer. Seasons play a big role in what weather extremes we might face at any time of the year.

Where sunlight is plentiful, solar panels can harness the energy and convert it to electricity. Extensive arrays of solar panels provide power to communities. Rooftop solar water heaters directly use the heat from the sun to provide hot water to homes and businesses.

The rays from our nearest star brighten our moods and brighten the landscape with glorious sunshine. Without that great big ball of fire, we could not see, there would be no weather, and we would miss out on the helpful vitamin D that sunlight helps our bodies generate. The sun gives off electromagnetic radiation. That might sound scary until you break it down. The heat we and the planet feel is infrared (IR) radiation. Light is visible radiation. Those two are good, when not in excessive amounts. The part of the sun's energy that causes tanning is ultraviolet (UV) radiation.

In summer, the sun's radiation is most intense since the sun is highest in the sky. As much as we love the sun, too much of that love will lead to pain. Over-exposing yourself to excessive UV radiation is not safe or healthy. In large doses, UV radiation causes sunburns. Sunburns develop from staying outside too long in direct sun, or from spending too many total minutes outside in a single day, even if there are clouds dimming the sun.

Over years, too much sun exposure dries skin, prematurely ages it, may lead to skin cancer like melanoma, and may increase the risk of cataracts. These are serious health concerns for all ages and all people. All things being equal, the degree of UV radiation harm depends on your skin pigmentation. The lighter your complexion, the more at risk you are of melanoma, but ALL skin tones are subject to skin damage. If you see anything unusual developing on your skin, have your doctor check it out.

UV radiation impacts are magnified by reflections in sand and snow. Crisp cumulus clouds may provide brief shade, but they also reflect radiation. UV radiation is stronger in mountains than at sea level because there is less atmosphere above to reduce the intensity. Even on cloudy days, UV radiation penetrates clouds to cause tans. Sunscreen is a must, and it has to be reapplied as it washes away by water or sweat.

Take sun safety seriously for yourself and your children. When your shadow is shorter than you are tall, radiation threats are high. Wear light-colored, lightweight clothing, along with a wide-brimmed hat. Use sunglasses that are rated for UV protection. Use sunscreen with a high sun protection factor (SPF). Be sure to follow the instructions for how and how often to apply it. Find shade under trees, canopies, tents, umbrellas, and overhangs. Shade will also reduce the direct heat your body absorbs from infrared radiation, so you won't feel as hot. Shade won't make a difference for oppressive humidity, though.

Get more sun safety and health tips through an online search on UV radiation from experts like the American Academy of Dermatology, the World Meteorological Organization, the American Medical Association, and the Environmental Protection Agency.

Around the United States, the National Weather Service and the Environmental Protection Agency issue an ultraviolet radiation index (UV index). The UV index describes the daily local strength of the sun's ultraviolet radiation. Some TV stations, newspapers, weather apps, websites, and radio stations put the UV number or category in the weather forecasts. The scale ranges from 1 to 15. The higher the number, the more intense and dangerous the UV radiation is. That means that your skin will tan and burn in a shorter time.

Each of us is affected differently by UV radiation, depending on skin color and type as well as our genetics, so use the scale as a guide. You have to make adjustments depending on how much your skin is affected by sun. With a low UV index, you still should wear sunscreen. A medium UV index is where you would be wise to wear a hat and sunglasses and limit your outdoor time. A higher UV index says limit your time in direct sun, but wear a sunscreen with a maximum SPF, while the highest UV levels tell you that your skin may burn in as little as ten minutes. Take all eye and skin precautions, and certainly limit direct sun exposure.

Even when the index is low, a simple cotton shirt blocks UV radiation, while shirts and clothing with tighter weaves are more protective. Sunglasses with UV protection along with wide-brimmed hats are basic tools. It should be obvious that sunbathing is not recommended as a healthy activity.

The sun is white-hot to our eyes. We can't and shouldn't look directly at it when it is high in the sky.

However, light from the sun lets us appreciate color. It highlights the bright white of midday clouds. As the sun lowers in the sky, we see other colors of the spectrum like yellow, orange and red. Sunlight passes through more atmosphere when the sun is close to the horizon, so colors with longer wavelengths become dominant while colors with shorter wavelengths are scattered out. This is why sunrises and sunsets often present dazzling and brilliant sky and cloud colors. Just about any color imaginable can be seen in the sky as sunlight interplays with air and moisture.

The moon reflects light from the sun. It also takes on a tint when it rises and sets. The color of the moon or sun depends upon how clean the air is. The hue is shifted by dust, ash, soot, pollen, or other particles in the atmosphere, known as aerosols. Even when the sun sets and is below the horizon, the afterglow tints the sky and clouds. Sunlight is refracted by water droplets or ice crystals and is separated into the 7 colors of the spectrum—red, orange, yellow, green, blue, indigo, and violet. The acronym for the color spectrum is ROY G BIV.

Earth atmosphere

A lot goes on in Earth's remarkably thin atmosphere. If Earth were the size of a soccer ball, the atmosphere that we know would be as thin as a layer of plastic wrap. When you fly cross-country in a jet airplane, 6 miles above the ground, most of the atmosphere that creates weather is beneath you. It rapidly thins with height. Weather occurs in this lowest layer of the atmosphere called the troposphere. That's where there is rising and sinking motion due to uneven heating from the sun and from varying topography. The air in the troposphere is mostly nitrogen and oxygen with small amounts of argon, carbon dioxide, hydrogen, and other gases. Air also holds varying amounts of water vapor, vital to life and weather.

In the troposphere, the temperature generally gets colder with height, the amount of water decreases with height, and the air pressure decreases with height. There's less air as you climb in altitude so it is less dense and cannot hold heat very well. That is why the tops of tall mountains, even near the equator, remain cold with snow on them, all year long.

The one thing that does increase with height is the average wind. These winds above the ground are what steer weather systems around the Earth.

Sometimes temperature does increase with height. This is known as an inversion because the temperature trend is inverted from typical. Temperature inversions result in stable weather because the warmer air aloft limits rising motion or convection. Temperature inversions are not unusual after clear, calm nights with no wind. When temperature inversions persist for days, they can cause bad episodes of stagnant air and pollution.

Given the greater density of the gases and water vapor low in the troposphere, the largest amount of the sun's heat is trapped near the ground. Without the troposphere, planet Earth would be a much colder place, unsuitable for life as we know it.

About 30% of the radiation arriving from the sun is directly reflected back to space by clouds and the surface of the planet. For the other 70% of the radiation, most reaches the ground where it is absorbed as heat while a smaller portion is absorbed directly by air. That's not the end of the process though. All incoming solar radiation, also known as insolation, is re-emitted at longer wavelengths between air and ground and clouds and water. Ultimately, there is a balance between the amount of energy coming into Earth's atmosphere and the amount leaving. This holds the planet at a nearly steady temperature. Most of this balance occurs in the troposphere. Fossil fuel emissions are shifting the balance.

The three components of the balance of energy are radiation from the sun; conduction of heat absorbed at the ground to the air adjacent to the ground; and convection, or rising air currents in the atmosphere.

Beyond the troposphere is a layer of the atmosphere where there is not much rising and sinking motion. The air movement is mostly horizontal and stratified. This layer is called the stratosphere. There's very little oxygen there but there is ozone which absorbs radiation, to make the temperature increase with height. The layer of ozone in the stratosphere protects life below from high levels of ultraviolet radiation. Ozone is a form of oxygen composed of three oxygen molecules, but we cannot breathe it. The oxygen we breathe is composed of two oxygen molecules.

Ozone in the stratosphere covers the entire Earth like a blanket. We call that the ozone layer because ozone is just one layer in the atmosphere, almost the way a cake can have different layers. The ozone layer is more than 10 miles above the ground. When ozone is in the stratosphere, that is good because ozone blocks ultraviolet radiation from the sun, which is harmful to people.

Scientists know that in certain times of the year, some of the ozone layer over the North Pole and South Pole disappears. They call this the Ozone Hole even though it's not really a hole. It is when the blanket of ozone that covers the Earth gets thin in one large area, to let more ultraviolet radiation reach the ground. Satellites measure ozone and show us where the thin spots are. So far, the ozone hole mostly happens in places where not many people live, but it worries scientists because they are not totally sure what makes the ozone disappear. They do know that certain kinds of air pollution get in the way of ozone forming in the stratosphere. *A lot of that pollution comes from gases called chlorofluorocarbons or CFCs.* Fortunately, many countries have stopped using CFCs. That pollution by itself is bad, but when it thins the ozone layer it's even worse.

Air pollution can actually increase ozone that forms near the ground, where we live. When ozone is near the ground, that is not good because it makes it harder for us to breathe. Older people, young children, and people with breathing issues and asthma feel it first. A lot of ozone near the ground will even make your eyes dry and itchy. Even some plants don't grow well when ozone levels are high.

The pollution that makes ozone near the ground comes from factories, power plants, cars, trucks, buses, motorcycles, ships and even gas-powered leaf blowers and lawn mowers. Some of the ingredients of gasoline, paints, and cleaners also play a role in forming ozone. When weather is warm and sunny, and the winds are light, different kinds of pollution mix in the air. If you add more heat, sunlight, or pollution, the ozone gets worse. When the sun goes down and the temperature is lower, ozone is not as bad, but it doesn't all disappear.

Ozone in your zone?

"Hey man, what's up? You look a little beat," Elliot said, extending his hand to Butch. "I know it's not the heat, 'cause it's not too bad today." "Thanks a lot. But yeah, you're right," Butch replied. "My eyes are tired, and I just can't get a good breath of air. I've had lung problems since I was a kid, and now, when the ozone levels go up, my energy goes down." "Ozone?! I thought that stuff was way up in the sky. I saw something on PBS that said it blocks harmful radiation from the sun." "Right you are, young Elliot." Butch nodded. "The problem is that we humans help to make ozone whenever we use things with certain chemicals or run engines that burn fuel."

Elliot looked up and scratched his head. "Is this connected to that Ozone Alert thing that I hear every once in a while?" "Yup. The ozone that we make stays near the ground. It's what most people call smog. It builds up pretty fast on a hot sunny day when the winds are light. If the levels get high, an Ozone Alert or Action Day is declared. People are asked not to produce pollutants that can help ozone grow." "Butch, you're not saying that I can't do my barbecue?!" "Yeah, that's what I'm saying, but there's more. It helps people like me if other people don't use gasoline-powered lawn equipment and if they don't put gasoline in their cars in the heat of the day. I know sometimes these things just have to be done, but when you plan ahead you can cut down on them."

"Hold on a minute," Elliot interrupted, "How come I can't do that if everyone else and all these big companies around here make a lot more pollution than I do?" "Let me finish. I was about to say that everyone and every business can do a part by simply cutting down on energy use so that the power company produces less pollution. People can also reduce car and truck use—vehicles are some of the biggest culprits in the game of ozone. Take the train or the bus and you might save some time too." "Well, you know my car doesn't even run so I guess I'm okay there." "That's true, but what about that job you had painting houses last summer?" "What about it?" Elliot asked. "Paint thinners, oil paints and even charcoal lighter fluid help to form ozone. If we can lower the ozone levels, then we can all breathe easier. In the meantime, when the levels get high, I have to go inside a building with air conditioning to get cleaner air." Elliot paused a moment, looked at Butch and said, "Man, I didn't realize that it really gets to you like that." "Actually, I can handle it now, but it seems like the older I get, the more these things get to me. So now you know what's going on, help me out, my friend." "You can count on me, Butch. I'm in your zone."

In urban areas, ozone is even worse because there are more cars, more factories, and more people. Big cities are also warmer than small towns and warmer air leads to faster ozone formation. High ozone levels near the ground cause bad air quality. When air quality is low you can sometimes see ozone and other pollution in the air. It makes the sky hazy. When ozone turns the sky yellow or brown, we call it smog. Smog is the words smoke and fog combined. It is unhealthy.

A change in the weather can clear the air. Wind brings in fresh air, but it also takes the polluted air and sends it somewhere else. Some of the air pollution rises into the clouds, where it mixes, to make even more pollution. When it rains, the raindrops carry the new pollution back to the ground, as acid rain. Acid rain can be so strong and dangerous that it becomes toxic, and slowly harms or kills trees and fish.

Remember that ozone in the stratosphere is natural and good because it protects us from ultraviolet radiation. Ozone near the ground is bad pollution. It makes it harder for some people to breathe.

Along with water vapor and gases, air pollution is part of what you see rising from exhaust stacks or smokestacks. Sometimes you'll notice dust, ash, and other tiny things that fall onto the ground or onto cars. These little particles can also make it harder for us to breathe because they go into our lungs and get in the way of oxygen. With high exposure over long periods, some of the particles remain in our lungs to cause damage over a lifetime.

It is very hard to live without making ozone and air pollution by mistake. The best thing we can do to make our air cleaner and healthier is to conserve energy, and not use as many fossil fuels. There are some fuels that make very little direct pollution. Three of them are a natural part of weather. They are solar power, water power, and wind. Think of how much power is generated by these ingredients in a thunderstorm. Unfortunately, there's no easy way to directly harness that. The size of thunderstorms tells you immediately that they are powerful. The tallest thunderstorms can reach the stratosphere, but when they encounter the stable air and horizontal wind of the stratosphere, around 12 miles up, they spread out and form a shape like an anvil. Because of the lack of moisture and air in the stratosphere, very little weather occurs there.

Two other layers exist beyond the stratosphere; the mesosphere and the thermosphere. These layers also protect us from high-energy radiation from the sun. The thermosphere is the edge of space. It is within the thermosphere where bursts of radiation interact with Earth's magnetic field to produce the aurora. High-energy particles travel from the sun in what's known as the solar wind. The particles constantly bombard our planet and are focused by the magnetic field around the North and South Poles. These are not directly dangerous to people on the ground, but when the sun emits large quantities of them, the particles can disrupt GPS and interfere with electrical power grids. They may force pilots to divert flights from over the poles for safety from excessive radiation. Most noticeably, this rush of particles across

Earth's magnetic field creates brilliant displays of the aurora, about 50 miles above the ground. A display of an aurora is space weather, a geomagnetic storm. In the northern hemisphere, it's called the aurora borealis or the northern lights.

Air

The troposphere's air serves as a blanket that allows life to exist. Air is mostly nitrogen, followed by oxygen. Nitrogen accounts for about 78% of air while oxygen accounts for around 21%. Hydrogen is only a fraction of a percent. There are much smaller quantities of other gases in air making up the remaining amount. The one gas that varies the most is water vapor. Water vapor can be as much as 4% of air. Everything directly or indirectly benefits from water and water vapor in the air. Plants benefit from nitrogen and humans and animals benefit from oxygen.

Air is an omnipresent vagabond. It's a gas that flows, moves, changes shape, and expands and contracts. While energy from the sun moves air, air in turn moves heat and moisture around the Earth. Air allows the planet to maintain a fairly consistent average temperature, affording large areas of comfy cozy conditions for humanity. Air contains natural and man-made particles and compounds known as aerosols and these travel within air. Some are irritants or pollutants, depending upon where they are found. They may move hundreds or thousands of miles across the planet. Particles or aerosols in air can change sky color, but in clean air we see a deep-blue sky. Blue is the color scattered the most by molecules in air. Without air, our sky would be black, as it is on the moon.

The amount of air is not the same everywhere on Earth. It changes daily. Where air piles up over a region, it creates high pressure and results in generally calm weather and light wind. That's the H you see on a weather map. The opposite of that is low pressure, where rising air creates clouds and unsettled and sometimes stormy weather. On a weather map, low pressure is shown as the letter L. Low pressure systems are also called cyclones, while high pressure systems are known as anticyclones. The lines of equal pressure sometimes drawn around highs and lows are isobars.

Air is not just outdoors, it is in buildings, in the ground, and in water. Without air, neither birds nor aircraft could fly. Air allows us to smell things and hear things. Without air, we would not be able to live on Earth. When air is still, we give it little notice; but the faster it moves as wind, the more powerful and noticeable it becomes. Air most often gains motion by the fact that it expands as it warms, becomes less dense, and therefore rises. That's a principle we clearly see in hot air balloons. Heat from the burners in a balloon warms the air and makes the air expand. The expansion of air enlarges the balloon as the air becomes lighter or less dense than the air outside the balloon. The balloon rises until it reaches air of equal density.

Rising air creates a region of less air or low pressure where it started. Air from somewhere else begins moving toward and converging upon that region of low pressure to fill it in. In a chain reaction, the original rising air expands, diverges horizontally and heads for regions of lower pressure. That air above the ground will sink when it ends up above where other air is departing sideways near the ground. The sinking air is compressed on the way down and piles up to create high pressure. High pressure and low pressure go together. This cycle never ends.

If you are a frequent flier, then you've been a frequent participant in an aircraft experiment known as takeoff and landing. Have you noticed that the altitude of the airport along with the air temperature controls how long you roll down the runway for a takeoff, and how long it takes the plane to stop when landing? At higher altitude, air is less dense. With high heat, air is less dense. Both of these create "thinner air" resulting in less resistance to any object moving through it. Less resistance is great if you are cruising along, but if you are trying to get off the ground or come to a complete stop, you want resistance! High altitude and hot air add to the amount of runway a plane needs to take off and land. It's one of the reasons why runways in mountainous locations and/or regions that get very hot in summer have long runways.

Wind

Air in motion is wind. You know it when you feel it. Wind may be light, strong, or extreme. Wind efficiently transports dust, pollen, seeds, moisture, and heat. This is nature's way of balancing components of our atmosphere. Moving air is noticeable in tree leaves, flags, smoke, and sometimes our hair. Wind creates the ripples on a lake and the waves in the ocean. Wind is invisible but it has power. What can wind do? The answer, my friend, is blowing in the wind. Wind is capable of turning turbines to generate electricity. Wind transports all the clouds. High wind may tumble or knock heavy objects over. Extreme wind can pick up and move large objects, including vehicles, in the case of a tornado.

Daily wind has a cycle. Away from coastlines, wind around sunrise is light or calm. It routinely increases in the morning and diminishes after the sun goes down, unless a storm is passing through. However, the wind a few miles up in the air is non-stop. When heat from the sun reaches the ground in the morning, the ground warms and then warms the air above it. That air rises. When the rising air encounters sideways moving air aloft, turbulence results and eddies of air with sideways momentum are created and carried down to the ground. Little by little, that wind, or momentum, works its way downward so that by mid-morning we feel wind. After sunset, the rising air slows down and stops, so the eddies or mini circulations also stop, and the wind diminishes again at the ground.

Wind is regularly stronger and more persistent along coastlines both day and night. In the warm seasons, the daytime difference in temperature between hotter land and cooler water creates rising air over land, resulting in a breeze that moves from water to land. This is called a sea breeze. Along a large lake, it's known as a lake breeze. The long travel over water without obstacles allows wind to be stronger on islands and along shorelines. This travel distance over water is known as fetch. At night, the breeze may reverse direction if the land cools off and gets colder than the water. That would produce a land breeze, with air moving from land, downhill and outward to the water.

Both of these local breezes are a type of circulation. Simply put, air movement takes on an almost circular motion with height. Air rises where it is warm. Aloft, it cools and spreads out to travel horizontally, then sinks to the ground to travel sideways to the area of rising air again. It's a wind circulation driven by differences in temperature. The temperature difference creates a difference in density, or weight of air, and that results in a circulation driven by heat, called a thermal circulation.

This same principle creates a circulation between the warm equator and the cold poles to transfer heat around the globe. In reality for the Earth, it's not one circulation but generally several that do the work of one. Each circulation is called a cell since it is independent from the others. For these large cells, air travels great distances over hours and days, so we have to account for the spinning motion of the Earth. The rotation of Earth on its axis causes a deflection of anything moving to the right of its path, in the northern hemisphere. The cells end up creating strong winds above the ground that blow eastward in the middle latitudes.

The strongest regular wind encircles the planet in middle latitudes around 6 miles above the ground. It's called the jet stream. The jet stream blows from west to east but it meanders north and south with dips, loops, and buckles. It's on the boundary of colder air from the Arctic and warm air from the equator. The jet stream wind gets its strength from a difference in temperature; so, in the summer when the Arctic has continuous daylight and warms, the jet stream is weaker and it shifts more toward the pole. In the winter when the Arctic cools, the jet stream is stronger, with winds sometimes over 200 mph, and it dips farther south toward the equator. It guides the tracks of large storms around the planet.

The jet stream is formed as warm air from the equator rises and goes north toward the pole. Because Earth is spinning, the northward moving air is deflected to the right. This is called the Coriolis Effect, named for a French scientist in the 1800s. It's the Coriolis Effect that gives the jet stream and our main steering winds a west to east motion. This west to east wind flow makes airplane travel from California to New York much faster than in the other direction, where you fly against the wind. The main jet stream winds are strong also because air traveling from the equator toward the pole maintains a higher angular momentum.

Objects moving across the northern hemisphere are always deflected to the right of their path. This is why air doesn't move in a straight line from high pressure to low pressure. Think of an old-school record player for vinyl albums. If you take an album that's spinning, and try to quickly draw a straight line on it with a marker, you'll see that the line always curves, no matter which way you draw it, simply because the album is rotating. Air leaving high pressure begins to turn to the right, or anticyclonically, because of the Coriolis Effect, also known as the Coriolis force. Air spirals outward from high pressure in a clockwise direction. Once it gets within the region of low pressure, the pressure is low enough that it pulls it inward, just like water spiraling down a drain, and then the wind blows counterclockwise or cyclonically, overcoming the Coriolis force. The greater the difference in pressure between the high and the low, the stronger the wind is. If you think of the H as a mountain of air and the

L as a valley, then something moving down the mountain will go faster if the difference between the peak and the valley is greater. That difference is drawn as pressure lines on a weather map. When they are packed together, it is known as a gradient, similar to the gradient on a topographic map showing how steep a hill is. A steep gradient of pressure means high winds.

Wind is also created when gravity pulls dense air downhill or through a canyon.

Do you have any idea which way the wind is blowing today? Do you care? You may think that you don't care, but you do. We all care because the wind brings changes to weather. When the temperature rises at night or falls in the middle of the day it's often because of wind. When the humidity jumps up and we see high heat index levels, you can credit or blame the wind. A shift in the wind can turn sweater weather to sweating weather.

Here's a trick question: Which way does a north wind blow? The answer is from the north to the south. Winds are named for where they come from. Think of it as a relative. If you have a cousin who lives in Georgia, in order for her to get to Michigan she travels to the north. When someone hears her ask, "How y'all doin'?" they'll probably ask her, "Where are you from?" She'll answer, "The South."

When the weather forecast calls for southeasterly winds in the summer, the Gulf coast and Eastern seaboard expect warmer and more humid air to visit because it comes from a region where the air is warm and humid. In contrast, north of the U.S.-Canadian border, air is usually drier and cooler so that a forecast for north winds means more comfortable summer weather will slide into the central U.S. from the north.

To figure out wind direction just look at which way flags are pointing or which way smoke blows. Watch the birds, who sit facing the wind. Look at a wind vane on top of some buildings. Your nose knows another way to figure out wind direction—your sense of smell and moisture. If you live close to an industrial complex, landfill, or cattle ranch then you really know what I mean. When wind blows from you toward the source of an odor (or aroma), you don't smell it much, but when wind shifts and you are downwind of something with a unique smell, you notice!

Depending on where you live, you can determine wind direction from what you hear and how loud it sounds. Think of all the things you hear on some days but not on other days, like distant trains, church bells, airplanes taking off, kids at a playground, and even traffic on a highway. Sound is carried farther by wind, so if you are downwind from the source, it's louder than it is for people who are upwind.

For large regions, the average wind can shift based on seasons, topography, and the proximity to large bodies of water. In the summer in the southwestern United States, the land stays hot and creates a persistent region of low pressure with rising air. Moisture from the Gulf of California is drawn inland and creates a pattern of almost daily heavy rain. This pattern is called a monsoon. That's not to be confused with a typhoon which is the same as a hurricane.

Some seasonal and local winds have unique names. A Chinook is a warm, dry wind in the western United States that blows down the slopes of mountains. As the air descends, it warms, and the relative humidity falls dramatically. The Chinook wind name derives from the American Indian tribe that inhabited the northwestern U.S.

In southern California, you find a wind called the Santa Ana. It picks up speed as it squeezes through mountain passes. The speed and gustiness of this hot, dry wind make it extremely dangerous in creating and spreading wildfires. This wind is named for the Santa Ana Valley region in which it is common.

Wind on a hot day is useful for helping cool your body. On a cold day, wind can rob you of heat. Any time the wind blows, it can be harnessed to do work in moving a sailboat, for example, or in moving turbines to create electricity. Wind power is endless. Wind never ceases. As long as the sun shines, there will be wind. As long as the wind blows, there will be weather.

Kitchen chef

Air, heat, moisture—the principles of weather are found in the kitchen. Try preparing a meal without those and see how far you get. Think of the chef who prepares a feast. Give the chef applause for applying principles of meteorology to conjure concoctions, and create a delight for your sight, and a treat for your tummy. A chef is a chemist, mixing solid, liquid, and gas. Gas could be natural gas, although air is the ever-present universal gas.

Heat, for meal preparation, is fundamental in the kitchen. It's heat that makes raw meats, poultry, and fish, safe to eat, and tastier. Heat has magical transformative properties. It can melt or meld, dry or fry, churn or burn. With heat, eggs go one way, while butter goes another. Moisture or liquid flows from there.

Have you ever made rice or mashed potatoes without water? That would be a drought leading to a dusty bowl. Count how many recipes call for some sort of liquid. Maybe it's quicker to think of how many recipes don't call for liquid. If the chef backtracks to all initial ingredients, he or she realizes that without water, there would be no ingredients. In the cooking process, water is in the beginning, and it's in the end when it's time to cleanup.

The chef can make it rain. Kitchen condensation is the water droplets that form on the lid of what the chef cooks. When it's cold enough outside, moisture forms dew on windows. Even at the table, your chilled beverage shows condensation on the outside of the glass.

We take air for granted, but the chef in Denver, Colorado, has to make cooking adjustments that the chef in downtown Miami doesn't make. Denver is a mile above sea-level, so the overall air pressure is about 15% lower than it is in Miami. The lower air pressure in Denver allows water to boil at a lower temperature. It's easier for molecules to escape the liquid to become a gas. Any recipe that calls for boiling has to be adjusted for high altitudes. On the other hand, the chef who uses a pressure cooker applies the principle that water under higher pressure will boil at a higher temperature. That makes pressure cookers good for faster boiling and cooking, at any altitude.

Don't distract the chef, though. The last phenomenon you want to see is a cloud of smoke, from too much heat, or not enough moisture, or both of the above.

CHAPTER 2

Water Cycle and Hydrology

What do two oceans do when they meet?
They wave.

Water falls from the sky in the science of meteorology. In fact, the word meteorology is the study of hydrometeors, aka precipitation. All forms of precipitation are hydrometeors. It's an incredible feat of physics that something so heavy is transported by air around the globe. These laws of nature are handled by the law firm of Precipitation, Evaporation and Condensation. Fresh water, H2O, enables life. Water is soothing. In many cultures, it is spiritual. It cleanses our environment and our homes. It's in every food we eat. Every creature needs water to survive. It makes up the majority of our bodies. Earth has a great deal of it, in oceans covering more than two thirds of Earth. The salty seas are just one place where water is stored. Fresh water is found not just in lakes and rivers, but in glaciers and in clouds. Water is stored underground in aquifers. Water floats as vapor in air and sits in snow and ice. It constantly changes phase and is recycled through weather in the water cycle or hydrologic cycle.

37

Start with water in the oceans and lakes and rivers. Heat from the sun causes it to evaporate. When water evaporates, it still weighs the same amount, just spread out over a much, much, much larger volume! That gives you an idea that what is in air is very heavy, including the air, itself. When water evaporates, it retains all of the energy that was put into it to change phase from liquid to vapor. This stored energy is known as latent heat. Wind redistributes water vapor around the planet. When the air becomes saturated with vapor, the vapor condenses into clouds and releases the latent heat, warming the air. Water is highly efficient at holding and transporting thermal energy.

Water vapor also condenses onto objects, often at night, when winds are calm and the sky is clear. Water droplets appear as dew on grass, cars and other things that cool down fast. The temperature at which this happens is the dew point temperature. You can tell dew is not rain because the sky is typically clear when you see dew, and pavement is generally dry. It's science and art when beads of dew adorn plants and flowers and even spider webs. When dew is heavy it drips off to the ground, and when the air warms, dew evaporates.

If the air temperature is below freezing, then water vapor forms ice crystals on those objects and that's called frost. Frost indicates that the temperature right at the ground or on a surface is at or below freezing. Window frost delivers delight in dissecting the delicate and detailed designs that develop. It's an artistic manifestation of meteorology. Frost on windowpanes or outside is the same as frost inside a freezer. It's water vapor transformed directly to ice without changing into water droplets. That's called deposition. The temperature at which ice crystals form is the dew point, but also called a frost point.

When temperatures are below freezing there won't always be frost. There's no frost when there's not enough water vapor in the air, when the frost point is very low. On the other hand, it is possible to get frost when the air temperature is above freezing because the temperature right at the ground or on surfaces is freezing or below. That's because colder air settles to the ground, and metal objects, like vehicles, radiate heat away, and cool faster than soil does.

Frost and dew don't fall from the sky, they form right in place when water vapor changes phase.

Tiny droplets of water in air, in large mass, form clouds. Clouds can also be made of ice crystals, depending upon the air temperature. Some clouds are a mix of both water droplets and ice crystals. Clouds are an effective vehicle for transporting fresh water around the planet. When cloud droplets or ice crystals grow and merge to become large and heavy, they fall as rain, snow, sleet, or hail, depending upon the temperature. Any form of water that falls from the sky is precipitation. When precipitation is heavy, we see and feel the incredible amount of water that can float in the air in both weight and in volume.

Much of the world knows rain, the pitter patter of raindrops on rooftops, at least in the warm season. You can sit back and listen to the rhythm of the falling rain. When rain hits the ground, it takes many different paths. Some rain will accumulate in large puddles and sit in place. Other rain will run off downhill into a creek, river, or lake. Some rain will slowly percolate into the ground. Rain and melted snow generally do all of these things, depending on what type of surface they fall onto.

Rain that sits in puddles may rapidly evaporate once the rain ends, especially if the air and ground are warm. The warmer things are, the faster and more dramatic is the evaporation. Snow can also go directly from ice to vapor in the process of sublimation but that's a much slower process. It's happening in your freezer right now. Sublimation explains why ice cubes in a freezer eventually shrink and disappear.

Any water that percolates into soil will recharge groundwater and aquifers and may work its way into a river or lake. Moisture pulled from the ground in tree and plant roots and given off from their leaves is called transpiration. That puts water back into the air.

If you have a boo boo and spill a glass of water, sooner or later it all evaporates. After a summer rain shower, we see rapid evaporation from hot rooftops, pavement, and roads. Weather drives the water cycle as a never-ending process, with no single beginning and all parts happening simultaneously. Each form of water over large areas modifies weather and all weather modifies each form of water.

As a liquid, water flows, changes shape, and is capable of storing tremendous quantities of heat. Liquid water can erode rocks and carry sediment and other heavy objects.

As a gas, water vapor also changes shape, flows, and stores latent heat. It is far more efficient at holding additional heat from the sun than dry air is.

As a solid, ice is rigid. It can reflect sunlight to insulate the ground. Ice can shatter and erode rocks and boulders. Ice is less dense than liquid water, so it floats. Ice can dam rivers to create floods.

Moving water does work by turning wheels or by turning turbines to generate electricity in a hydroelectric plant. Water and the water cycle play a tremendous role for the planet.

Floods

Rain that coats roadways can hamper your scamper, as conditions go from drippy to slippy, with tires and shoes losing traction. Wet surfaces demand awareness. Floods require caution. Too much water that rises in a normally dry spot is a flood. Floods can cover small areas, like a road; medium areas, like a neighborhood or farm; or large areas, like a city. Flash flooding is the fastest type of rising water. Flash flooding is most common when heavy rain falls over the same spot and can't drain fast enough. The floods may come from intense thunderstorms sitting in place, or from several hours of heavy rain. As water always runs downhill and rises in the low areas within a watershed, drainage systems may be overwhelmed leading to street and neighborhood flash flooding. Creeks and rivers also may rise quickly from extreme runoff and then overflow onto nearby roads and property. Both of these may occur together. All flooding is worse when soil is saturated or when the ground is covered by impervious materials.

Floods could be standing water or moving water. Swiftly moving water, especially when it is carrying debris, has the energy to eat away hillsides and undermine the soil of roadways. In the worst cases, a road may be washed away beneath the floodwaters. For road foundations that are partially compromised, it may be hours, days, or weeks after extreme flooding before a road collapse occurs.

Flash flooding takes lives because people underestimate the power of moving water, even when the water seems to be shallow. Every cubic foot of water weighs 62 pounds. A vehicle crossing water that's not very deep can easily lose traction and be swept off a roadway into deeper water. A person walking through fast-moving water that's less than a foot deep can easily lose footing and be knocked down and carried away.

River flooding is a slower and more prolonged process. In regions where heavy snow is common, river flood threats are seasonal. The warmth of spring triggers the melting of a snowpack built up over the winter. When snowfall is far above average and then temperatures rise above freezing for long periods, a huge amount of meltwater will run off and enter the rivers. In extreme cases, the rivers increase in width, depth and in speed. They spill over their banks and flood low-lying land in floodplains for days or weeks. It's a natural cycle, but when roads are cut off and homes and communities are isolated or impacted, it's a disaster. Rivers with melting ice also may rise further and suddenly if the ice creates dams.

Any river causes erosion and may carry large amounts of sediment or debris. Slow erosion happens on time scales of decades and years in large rivers. In rapid flooding, erosion occurs over days and even shorter periods like hours in rivers, streams, and creeks. We see erosion in neighborhoods, not just from flooding, but from runoff that continually focuses moving water over one area.

Flooding increases the potential for destruction of property and disruption to society. Rising water that damages property is usually not covered in standard property insurance. Search online or ask your insurer, whether you live in a flood zone or floodplain, if separate flood insurance is required or suggested for your community.

A Flood Watch means flooding is possible and you should be prepared to take action. If your home is in a flood-prone area, you must always have a plan to protect property, and life. That includes having alternate escape routes if access or egress gets cut off. A Flood Warning means flooding is happening and you must take safety steps immediately.

With population growth and expansion, many homes and businesses are developed closer and closer to floodplains. A floodplain is the natural basin for river overflow. Floodplains are not normally wet, but they go through cycles of being dry and holding water. New development also can redirect and increase the flow of water runoff making water rise in places where it never flooded in the past. Development may hide the existence of floodplains, but when water rises, the floodplains reappear. These are things we must account for in long-term planning for the threat of floods.

Precipitation

Rain replenishes lakes, rivers, aquifers, and fresh water supplies. Rain cleans the air and makes it possible to raise livestock and grow crops. In extreme cases, lack of rain results in drought while too much rain is a hazard that creates flooding and erosion. Rain can be light, moderate, or heavy, depending on the type of cloud from which it falls. Heavy rain is common in big low-pressure systems like nor'easters, tropical storms and hurricanes, or even just from thunderstorms that move slowly.

Light and moderate rain is routine from stratus clouds found within broad regions of low pressure. Steady rain from stratus clouds tends to cover extensive areas since the clouds are wide and spread out. The lightest rain is called drizzle. Drizzle is tiny water droplets that slowly fall and dampen everything. Drizzle doesn't add up to much.

Rain that falls briefly and stops is called a shower. We get more showers from convective or cumulus-type clouds. Since these clouds can be small, it wouldn't be unusual to get a shower on one street and just a few blocks away the ground stays dry. Showers may be light, moderate, heavy, or intense. Snow also falls as showers from cumulus clouds.

It's fluffy, it's bright, it starts out sparkling white. It comes with the cold and turns dark when it's old. A child's playground, a traffic nightmare, it gets in our clothes, it gets in our hair. Some love it, some hate it, either way we can't mitigate it. "But what is it?" Snow!

Snow is flakes of six-sided ice crystals that fall from clouds. Take a moment to look at some flakes the next time it snows. Catch a few in your gloves or on a coat sleeve. Examine them closely without letting your hot breath melt them, and you may be amazed at the amount of precision and detail in each flake. Depending on the temperature and the type of weather pattern, snow can range from perfect hexagonal star-like flakes to clumps of randomly oriented blobs.

"How cold does it have to be before it can snow?" There's a short answer and a long answer to this. The short answer is 32 degrees Fahrenheit. Above 32 degrees you won't find ice. The long answer is the temperature must be 32 degrees or colder above the ground so that snowflakes fall without melting on the way down. Therefore, it can be warmer than 32 degrees near the ground as long as the air above is colder than 32 degrees in a deep enough layer or multiple layers.

"So, how warm can it be before you can't have snow?" That depends on how high the clouds are and how much colder the air above the ground is. There is no magic number. It wouldn't be unusual to see snowflakes, with near-ground temperatures in the lower 40s. Under the right circumstances of very cold air above the ground, one might see a flake or two with the ground temperature near 50 degrees. Of course, the snow would melt as soon as it hits the ground.

Aside from putting snowfall into the categories of light, moderate, and heavy, you can separate it into flurries, snow, and snow showers.

Flurries are snowflakes that you can see, that easily blow around, but they don't totally cover the ground. In other words, you can't measure them. Snow is a continuous period of snowfall that is pretty uniform over a large area, such as a city or county. Snow showers are similar to rain showers in that they are not steady, and they don't cover a wide area. With snow showers, some neighborhoods may get a couple of inches of snow while across town only flurries occur.

"How is snow measured?" The easy way is with a ruler. Just stick it in the snow in a few places to take an average depth. Meteorologists use rulers that are divided into tenths of an inch. On windy days or when the snow melts quickly, it's difficult to get an accurate measurement. Another way to measure snowfall is to catch it in a tube, melt it, measure the water as precipitation, and then convert that back to a snowfall.

"How do you convert between precipitation and snowfall?" The rule of thumb is ten inches of snow is equal to one inch of rain. This ratio of 10 to 1 is only a guide. It changes with the temperature. The colder the air is, the fluffier and lighter snow is, so the ratio might be 20 to 1 or greater. When air is closer to freezing, especially if it is a little warmer above the ground, snow will be wet and heavy, and the ratio might be 5 to 1 or less.

"Can it ever get too cold to snow?" No, it's never too cold for snow. Even though bitter cold air holds little moisture, once you see clouds there is always the possibility of snowfall.

"What's 'lake effect' snow?" Lake effect snow is common on the eastern and southern shores of the United States' Great Lakes in fall through winter. In the depths of a cold period when the lakes are not frozen, water evaporates from the lakes to form clouds over the water. Because the water is much warmer than the air, lots of moisture rises and saturates the air to the point that the clouds release snow. The clouds and snow showers are carried by wind to the shoreline where the snow falls as lake effect snow. It's not unusual to find streets near the lake filled with 5 to 10 inches of snow while the sun is shining 5 miles away. Lake effect snow can also occur along with a regular storm system that has cold winds blowing a long distance over the unfrozen lakes, making forecasting snow amounts a huge challenge.

Snow is the winter equivalent of rain, but snow is not simply frozen water. Snow is crystals of ice, forming directly from water vapor, which bind together to create flakes. A snowflake can be as small as a grain of sand, or as large as a button. A single snowflake may be made of hundreds of ice crystals. With ice crystals forming a lattice, snowflakes may be flat, columns, or needles. All branches occur at an angle of 60 degrees, based on the molecular structure of water. Snowflake sizes and designs depend on the amount of vapor available when they form and how the temperature varies in the cold air where they develop. Individual flakes may have a simple appearance or one that ranges from elaborate to ornate. No two snowflakes are truly the same. While most snowflakes are not perfectly symmetrical, they all start with 6 sides.

It's not unusual for a snowflake to have different types of ice crystals all stuck together. As they fall from miles up in the sky, the flakes tumble, break apart and merge, moving through air layers of differing humidity and temperature.

Snow has a huge impact on transportation. Large snowfall amounts make travel hazardous. Towns may spend hundreds of thousands of dollars each year to keep roads safe, and to allow commuting and commerce to continue. Big cities spend millions of dollars annually for snow removal.

Even on a sunny day, blowing snow from the ground that reduces visibility to under a quarter mile when the winds are over 35 mph for several hours is called a blizzard. The definition of a blizzard does not include temperature. It's only about wind-blown snow that makes it hard to see. Many people erroneously interchange blizzard with heavy snow, but a blizzard is more than heavy snow. It is also a whiteout that could last hours.

Snow gives good opportunity for recreation and exercise. Sledding down a hill usually brings smiles to many faces, as long as people are properly dressed, and the temperature is not too low. Cross-country skiing and downhill skiing are popular in many regions. They continue outside of winter, too, in high elevations where snow is slow to leave because the air stays cold. Snowmobiling is also another winter recreation, that equally serves as transportation. Cold winter air, for long periods, is definitely required for ice fishing. When lakes freeze, since ice is less dense than water, the ice floats and the water remains liquid below. That's why the fish survive.

Regular precipitation is critical for humans and other species to maintain health and adequate food sources. Some of the coldest places on Earth are actually deserts, like Antarctica. We think of deserts as dry hot places, but the definition of desert does not include heat. Even if snow and ice is on the ground, a location is a desert when very little precipitation falls. It's the lack of substantial precipitation over centuries that creates a desert. On the other hand, a persistent lack of regular precipitation over months and years leads to droughts. A drought is not just dry, it's much drier than average. A drought can be defined by how much below average the precipitation is; how arid the soil is; how little water is in the ground or aquifers; how much below demand available water is; or by combinations of any of these.

Rain and snow can fall together in any combination. Snow that falls and halfway melts before refreezing, or rain that halfway freezes before hitting the ground, is called sleet or ice pellets or graupel. When rain falls through air that is above freezing but then makes contact with a surface that is below freezing like pavement, vehicles, power lines, or trees, it freezes on contact. That is known as freezing rain or an ice storm. It may be slow, gentle, and silent, but an ice storm has great negative consequences. If you take the weight of water that coats trees and leaves, you can imagine how layers of ice building on any structure create extreme weight and stress. Ice storms can take down thousands of trees, and powerlines. Light rain or even just drizzle accumulating as layers of ice is an extreme hazard. Powerlines and trees often fall from the weight of ice. Ice storms can paralyze a city or a region. People may suffer without electricity and heat. Travel is nearly impossible.

Glaze or any thickness of ice closes streets and highways. All-wheel drive does not guarantee vehicle control. Trucks without chains lose traction. Their weight and power are no match for ice on an untreated roadway. One single accident on a major travel route can strand thousands of people for hours. When there is a glaze on roads that is hard to see, it's nicknamed black ice, because it blends with road surfaces. In regions not accustomed to winter precipitation, an ice storm, followed by days of subfreezing weather stops everything.

Hail is a different form of ice. It can happen in the warm season when the temperature near the ground is far, far above freezing. Hail comes from thunderstorms or clouds with a lot of rising air and ice aloft. Water is suspended in the sub-freezing air by rising air currents, miles above the ground. It freezes aloft in a process that builds layers and layers, like those of an onion. Hail may be a single ball or a chunk of ice, entirely composed of layers, or it may be dozens of smaller pieces of ice clumped together, or a combination of the two. When the ice gets too heavy to be supported by the upward wind, it falls to the ground. Even though hail may partially melt on the way down, at the ground, it can range in size from that of a pebble to that of a grapefruit. The larger the hail is, the stronger the upward wind or updraft is in the storm and the more intense the thunderstorm can be. For example, 1" hail or hail the diameter of a quarter is caused by updrafts over 50 mph. When hail is the size of a tennis ball, the updraft is at least 75 mph and for large hail—the size of a softball—the updrafts are over 100 mph.

For clouds that are high, precipitation may fall into a layer of dry air and evaporate before reaching the ground. That is called virga. The evaporation of precipitation causes air to cool, so oftentimes the wind becomes gusty beneath virga as colder, denser air sinks from the sky. In extreme cases, this downward rush of air creates dust storms, especially in arid regions. Intense dust storms that come on suddenly, with a dramatic, tall forward edge are known in the Middle East as haboobs. Haboob is an old term of Arabic origin, used in the Middle East to describe large dust storms. It's a word you'll hear more often in the U.S. desert Southwest because dust storms are frequent there. Dust storms can be far removed from the original precipitation and storm that caused them, because wind may travel for dozens of miles, picking up loose soil, sand, or dust.

Annual precipitation types and amounts are controlled by wind direction, proximity to large bodies of water, temperature, and topography. In general, warmer regions that have wind blowing from water to land, will have higher precipitation totals.

Rainiest city

I've lived in Mobile, Alabama, the birthplace of Mardi Gras, for decades. You may have heard it called the wettest city in the country. It was proclaimed such by an online magazine. In a rain Olympics, I suppose we would be awarded a golden galoshes medal. For Mobilians, it was something to brag about to family in Phoenix, but there are many caveats to being the wettest city. Like a legal contract, there's fine print. Here's how that statement needed to be phrased:

Based on the average annual rainfall at NOAA and weather observer reporting stations, in the 48 contiguous U.S. states, between 1981 and 2010, Mobile Regional Airport was the wettest spot within a populated city (not town), in records held by the National Centers for Environmental Information. The average yearly rainfall was 66.15 inches. Mobile was actually not the wettest city in each of those years. As data for what is considered average (or normal) is recalculated every 10 years, the wettest city title is one that was sure to change. Mobile was wettest only by a fraction of an inch, not by practical measures.

Mobile Regional Airport, where the official rain amount is measured, is not the wettest *location* in the 48 states. In that same time period, next to New Orleans, Algiers, LA, averaged slightly more rain than Mobile at 66.40" but it's a much smaller city. Pensacola, Florida, averaged just over 65" while Miami, Florida, was around 64". The Gulf Coast is wet and south Florida is wet. Mobile's rainiest title was a matter of semantics, how you define populated city, and how well-calibrated rain gauges are. Rainfall in Mobile is not dramatically higher than other Gulf Coast locations. Throw in a hurricane or tropical storm and any other spot can win the title but end up a loser.

"Doesn't it always rain in places like Seattle?" you ask. By perception, yes. Seattle's average yearly precipitation (including melted snow) is under 38," BUT there are over 150 days per year in Seattle where rainfall can be measured, even if it's only a tiny bit. Compare that to Mobile which "only" has rain on 120 days. Coastal Washington State claims many locations that get more days of rain, but the Gulf Coast gets more rain on fewer days.

In the coastal mountains of Washington and Oregon, where there are no communities, precipitation amounts can easily break 100" in a year. That includes melted snow. The National Park Service says Mt. Olympus, west of Seattle, gets 270" of precipitation per year. If you head to the 50th State, Hawaii, Mt. Waialeale in Kauai can get over 400" of rain annually!

Water resources

The cycling of water between air, land and sea as liquid, gas and solid, is critical to planet Earth and to life. Nobody owns the water on Earth. We share it, with a moral and practical obligation to conserve and protect it. Water ties together many lives and systems on Earth in ecosystems. In the global ecosystem, every living plant, tree, fish, insect, organism, animal, and human shares water. We use water to drink, prepare food, bathe, clean, and to stay healthy. We also use water for cleaning, recreation, shipping, industry, travel, and as a source of food, and income. The cotton in our clothes needs water to grow. Fruits and vegetables are filled with water. Water is used in printing, painting, pottery, and pouring cement! Moving water is used to make electricity. Water can cool a car engine or cool you down in the hot summer. It can make tremendous storms that carry rain across the oceans. Many of us in developed nations don't think much about water because it's always there, but if water supplies get low or compromised, we notice.

The total amount of water on Earth does not change. That's amazing when you think that the water in your faucet may have been deep in the sea millions of years ago. Or it may have been in a glacier for thousands of years.

When you are done with it, it goes down the drain and sooner or later back to a lake or ocean where it may rise up into the sky and form a cloud that travels around the world to be used again. Fresh, clean water is precious. If all of the water on Earth were poured into 100 large buckets, 97 of the buckets would be salty ocean water.

For human needs, fresh water is not available where we want it to be at all times. Some parts of the world are just dry, while other parts periodically go into a drought. A drought is not just about lack of precipitation, it's about supply and demand. Population growth and drier decades produce more demand than supply in many western areas of the United States. A good-hearted person suggested that the rainy U.S. Gulf Coast send water to the drought-stricken western U.S. The noble thought of sharing water brings up a huge bag of legal, economic, environmental, and ethical issues, though.

In some U.S. states, rainwater is owned by the state, and you cannot catch and retain the rain that falls on your roof in a rain barrel. You are to let it run off, naturally, to recharge aquifers. In some communities, you are not allowed to purposely direct rainwater away from your property, mostly to not overwhelm the drainage systems.

Who would pay for the capture, transport, and distribution of water from the Gulf Coast to California, just for example? A single cubic foot of water (8 gallons) is

62 pounds. That would be a tremendous amount of dollars. Maybe you could pay to build a pipeline but then you get into property rights, eminent domain, and aesthetics of a pipe. When the drought ends or shifts, who would pay to dismantle or move the pipeline? Gravity won't send the water away, so water would have to be pumped. From the Gulf Coast, most of the country is literally uphill. That means a need for power. A pipeline or parade of endless railcars for water transport would likely rely on fossil fuels to power them, and that would lead to pollution and carbon dioxide emissions.

What if all Gulf Coast property owners chose to retain the rainwater that falls from the sky on their property, and not contribute to a giveaway? Aside from legal fights, that would cause rivers to run low, and freshwater feeds to estuaries and bays to diminish. Ecosystems would be damaged. Subtracting large amounts of water from a wet place or adding it to dry places would change evaporation amounts, with some change to local weather.

What if a region gives water away, and then goes into its own drought? Do people ask for the water back? Should water be sent to a region where it is being used unwisely—for things that are not critical to life? Should it be sent to an area where the water would be used for agriculture to flourish, giving that region an economic boost, while maybe jeopardizing the sender's agriculture, fishing, and recreation sectors? Don't take water for granted. Look around the world where there is famine. When water is not available, people, animals, and communities suffer, often with economic disaster and political unrest.

We think of water as liquid, but when water freezes, it changes phase and becomes ice. When ice accumulates from snowfall and does not melt for hundreds of years it's a glacier. There is more fresh water locked up in the ice of glaciers than you find in all of the lakes of the world, but it is still a small amount compared to the salt water in oceans.

When we drain ponds and wetlands to build farms or new communities, we change ecosystems and habitats. When rivers are blocked to create lakes, animals and people are displaced while the lives of fish and other creatures are changed. If one farm uses too much water or allows pollutants to run off into a river, everyone is affected.

Did you know you are in a watershed right now? Everybody lives in a watershed. A lake is an obvious watershed, but a watershed, in general, is an area where all the water drains to one spot. Precipitation will run off and drain into creeks and rivers that are all in a watershed. Many large watersheds end in estuaries, where fresh water enters the oceans and mixes with salt water. Estuaries are home to a wide variety of animals, fish, insects, and plants that depend on good water quality. Any place inside the watershed, you may also have areas where the ground is sometimes wet, and other times dry. That's a wetland. Both estuaries and wetlands are very important in having an ecosystem with healthy fish and animals, and clean water. Watersheds, estuaries, and wetlands are impacted by weather, and they impact weather.

Since all water is shared on Earth, if it gets polluted it can lead to crisis. Seemingly clear water can be polluted if there are chemicals, insecticides, fertilizers, or toxins in it. Water runoff from a heavy rain can be too warm for fish in a river and that's a kind of thermal pollution. Water pollution happens whenever the water is not healthy.

Ships sometimes spill fuel or oil and that kills fish, compromises water quality, and may cost millions of dollars to clean up. In some cities, there are factories that unintentionally or carelessly put oil or chemicals into the water. Even exhaust stacks or smokestacks make pollution that rises into the sky and is washed back to the ground in rain. That rain is sometimes called acid rain and it harms plants, and the animals that eat the plants.

When a factory puts a lot of pollution or waste into rivers or lakes, that is known as point source pollution, but factories are not the only polluters on the planet. You might guess that farms make a lot of pollution from fertilizers, pesticides, or waste from animals that washes away. Some may, but farms are not the only polluters either. Water polluters are you and I, and our friends and families. We generally don't do it on purpose, although there are people who litter carelessly, and dump substances intentionally and illegally, but pollution happens a lot because there are so many of us, scattered all over the world. The pollution we produce is called non-point source pollution because it doesn't start at just one point.

Our cars and trucks leak oil that gets washed away. Boat engines also can leak oil. When we add excess fertilizer or weed killer to our lawns, that is washed away too. Some people pour old paint and chemicals down roadway drains and that goes to the rivers. Everything that washes away goes into a drain, then to a river. It disappears from where we see it, but it always goes somewhere else. Even if we don't pollute water, we create wastewater. It's impossible to live and not make wastewater and pollution. We just need to try to not make so much. Many things we put into water will slowly break up, but they are still there, on some level.

There are other things we do that waste perfectly clean water. Some of us leave the water running while we brush our teeth. Others of us spray our lawns on windy days when the wind carries the water away. Others let the water run for a long time, waiting for it to warm up or cool down. Leaky pipes waste water too.

There are many ways you and I can help to conserve precious water and reduce water pollution:

- Clean up litter so that it does not get washed into rivers and lakes.
- Save poisonous or hazardous chemicals and take them to a collection center.
- Use natural items like baking soda or vinegar to clean, instead of a chemical cleaner.
- Turn downspouts so that the water soaks into the ground instead of running off onto pavement.
- Plant trees and bushes around your home that don't need extra watering.
- Use a car wash that recycles water.
- Recycle metal, paper, plastic, and glass, so that factories don't have to use water to create more of them.
- If you have a dishwasher, don't run it until it is full.
- Shower instead of taking baths.
- Install showers and faucets that use less water.
- Direct condensate from your air conditioner to a flower garden or part of the yard that needs water.
- Make sure there are no leaky pipes where you live.
- Reuse water from the sink for watering plants or rinsing dirty dishes.
- Plant grass or ground cover to keep soil from washing into creeks and streams.

- Clean up after your pet and flush the waste down the toilet.
- Sweep up litter and dirt on your sidewalk instead of spraying it away with a hose.
- Make a compost pile to recycle leaves, grass clippings, and fruits and vegetables.
- Put compost around bushes and plants to keep the soil from losing moisture.
- Ride your bike or walk for short trips instead of riding in a car.
- Test the water quality in local streams and lakes.
- Adopt a local stream and help to keep it clean.
- Organize a lake or beach cleanup to protect the fish and animals.
- Before you go swimming, use the restroom.
- Use silt fences or barriers to block sand and soil from washing away.

Nature does an incredible job of cleaning water in the ground, and in the air when it changes phase from liquid to water vapor to ice crystals. A growing population with increasing resource usage challenges that. The amount of water on Earth does not change and that means as more and more people use it, we have to plan better. We also have to respect the power of water when we are around it, and during dangerous floods. Learning how to swim and knowing water safety rules will make you safer while having fun.

Water is all around us, above our heads, under our feet, in our bodies, and in the foods we eat. Meteorology is everywhere, even in a grocery store.

Cold front in aisle eight

"Will that be paper or plastic, sir?" the grocery store bagger asked, rolling up his sleeves. His name was Kyle. "Uh ..." as I realized he was addressing me, "Plastic. It's easier to carry." I watched the price readouts as the checker scanned my items. Between beeps, I stole glances toward Kyle bagging my groceries. He was doing something that not enough baggers do, separating cold items from the others. "This young man has a good future," I thought, "He's polite and he knows a few basic principles of meteorology."

The checker swiped the frost-covered frozen orange juice over the scanner. "Kyle, is this a two-fer-one?" "It was yesterday. Le'me go check." He squeezed between the cart and rack of tabloids and ran toward the frozen food section. Most of the supermarket was climate-controlled to a dry 68 degrees. However, some areas were different. Along aisle 1, where the fresh fruits and vegetables were, the temperature was lower. Every five minutes the automatic mist sprayers triggered like a spring shower to help to maintain a moderate level of humidity. The bakery was almost tropical. The continuous cooking made it warm and humid. In aisle 7, the cereals and baking goods were found in a climate of low humidity. Moisture does bad things to powders and mixes. What's worse than opening a bag of sugar and finding chunks rather than granules?

As Kyle rounded the bakery to turn into aisle 8, he skillfully dodged two girls playing unattended with shopping carts. He traveled down the parallel rows of polished steel and glass display cases containing frozen pizza and microwaveable dinners. Orange juice was in a bin at the end of the aisle. Meanwhile the two children had started to open the doors to the ice cream freezer. They were about to get a lesson in meteorology. As the older girl held open a door her sister said with amazement, "Look! The glass is fogging up." The younger sister started to draw pictures on the glass as her sister decided to see how many doors she could open and fog up. "Look at the mist coming out of the bottom of the freezer," she said with a giggle. "I'm getting goose bumps."

Kyle turned to retrace his steps to the register. Before he could ask the girls not to play with the doors, their father appeared. "What are you two doing?" he sternly asked. Quickly shutting the doors, the girls replied in unison, "Nothing." Kyle smiled and kept on moving.

Once back at his post, he continued bagging duty. The juice was two-for-one. I got a deal that day.

When you purchase groceries, be like Kyle, don't pack wet with dry, or cold with warm. Not only is food less appetizing when it's at the wrong temperature but it won't last as long in storage. Think humidity and condensation when bagging. All air has some water vapor in it. Even air inside packaged dry goods such as bread, cookies, cereal, or rice is not totally dry. When air is cooled to the point that it can hold no more water vapor, you'll get condensation or dew; that's the dew point. This is why when you take a cold beverage from a refrigerator and set it on a counter the glass may "sweat." The "sweat" is moisture that was in the air that condensed on the glass. Since frozen items are obviously cold, a lot of water will condense on and around them once they are out of a freezer. Therefore, if you put frozen juice next to a loaf of bread you get moisture condensing inside the bag leading to soggy bread. The moisture was already inside the bag. The cold orange juice caused the temperature to fall within the bag until the air was saturated, at its dew point.

Some stores place frozen items in a plastic bag within the grocery bag. This is not just to catch condensed moisture. The inner bag provides a layer of insulation between the cold items and the warmer air outside, so there will be less condensation overall. And what about the two girls? After a near-spanking, their lesson in meteorology was completed when their dad correctly explained that when moisture from outside air gets into the freezers, you'll get condensation inside the cold glass doors. Cold air is more dense than warm air, so it sinks to the floor like a fog. The girls are smarter now.

CHAPTER 3

Clouds

What cloud is so lazy it won't get up?
Fog.

Clouds are as old as the Earth, serving as inspiration for many a poet, artist, and meteorologist. As magnificent masses of moisture adorning the sky, many clouds are similar; however, no two are ever the same. There are endless varieties of clouds that overlap in time and space. Clouds are like people—they come in different sizes, shapes, heights, and colors, but they are all made of the same thing—water. If we were to give clouds personalities, some would be introverted and barely noticed, while others would be extroverted, commanding your attention. A good meteorologist scans the sky for clues of change. When clouds abound, the type, quantity, and motion tell the story of weather. Even a non-meteorologist can instinctively identify clouds of a peaceful pattern or clouds of chaos.

Names for clouds have Latin origins, describing a genus and species, with more names for supplemental or accessory clouds or features. Most of the current names have been in use by scientists for over 200 years. Proposed and promoted by Luke Howard, the naming system continues to grow as we categorize more variations of clouds.

There are two general classes of clouds: those that bubble upward to become tall; and those that stretch out and are wider than they are tall. Clouds that grow upward are called cumulus. They accumulate in one spot and swell. They can resemble cotton balls or cauliflower. Cumulus clouds are the ones most of us learned to draw pictures of as children. Cumulus clouds indicate vertical motion in the atmosphere, called convection. They are common where humidity and warm air are abundant. Sunlight heating the ground creates plumes of heat and moisture that move straight up to create cumulus. Air that is pushed against mountains and forced to rise can produce cumulus clouds too. As rising air cools, the relative humidity rises to 100% and the water vapor condenses to form bubbly clouds.

There are many varieties or species of cumulus. The smallest are fair-weather clouds that tell us things should remain calm or fair. They are classified as cumulus humilis which translates to humble accumulation. They all are about the same size and the same height above the ground, spaced far apart, bright white and set against a blue sky.

Medium-sized cumulus clouds are called cumulus mediocris. These clouds are not weather-producing clouds when they float in flocks with equal size. Cumulus mediocris are just thick enough so that daylight may not penetrate from the top to the bottom, leaving the bottom darker. That makes it more noticeable that they have flat bases. The height of the cloud base tells at what altitude the relative humidity in rising air reaches 100%.

Medium or small cumulus clouds allow a good deal of sun to reach the ground. Their shadows show the cloud size and speed of motion. When cumulus clouds grow large, they become cumulus congestus which means congested accumulation. This is when showers of rain or snow become likely, depending on the air temperature.

The largest cumulus clouds are the ones that produce rain, lightning, and thunder. The name for these is cumulonimbus. Any cloud that produces precipitation is of the nimbus variety. Large cumulonimbus can block sunlight and rise 10 to 12 miles high. They contain an enormous amount of energy. The upper portion of the cumulonimbus is typically ice and water, even in hot regions and seasons. Cumulonimbus clouds are capable of generating violent weather, from lightning to intense rainfall with flooding, to hail, to extreme wind, to tornadoes. Cumulonimbus can drop snow in cold regions, in the winter.

Just as in music where you hear tenor, alto, and soprano singers representing different pitches or ranges, clouds are organized into different heights with alto meaning middle, and cirro meaning highest. Also similar to a singer who might be strongest in one range, but can overlap into neighboring ranges, clouds with a designated height can overlap into other altitudes. Altocumulus clouds are a good example of that.

Altocumulus is a gentler type of cumulus cloud that forms in middle levels of the troposphere, but is sometimes found at higher levels or lower levels. Altocumulus clouds are often layers, patches, or sheets showing rounded masses or rolls that may be medium or large. Sometimes resembling curdled milk or the scales of a mackerel fish, they are made of either water droplets or ice crystals, depending on the air temperature. Altocumulus clouds of water droplets create a bright area of colored light around the moon called a corona.

At the highest levels of the troposphere, cumulus clouds exist as cirrocumulus and they are made of ice crystals, rather than water droplets. Cirrocumulus clouds have tiny elements like ripples or grains that do little more than dim the sun or moon. Cirrocumulus and altocumulus, by themselves, are usually associated with calm weather.

Cumulus clouds that grow fast within moist air may have a layer of cloud just above them called pileus. It's also called a cap cloud because it looks like a cap. Pileus clouds form when rapidly rising air forces a moist layer of air just above a cumulus cloud to gently rise, cool, and then condense into a smooth cloud. Pileus typically are brief, lasting a few minutes.

Partly or mostly

Martin slouched back in the chair and patted his stomach. "That turkey hit the spot," he said to Miranda. "I'm quite the chef, if I must say so myself." Miranda rolled her eyes and smiled. "Well, Mr. Chef, do you want any more of that apple juice?" "Uh, yeah, but just half a glass." "Which half?" Miranda asked slyly. "Do you want it to be half full or half empty?" Martin chuckled. "You've been reading those books again, haven't you?" Miranda poured the juice into the glass. "Actually no, but I started thinking about perspective when I was listening to the weather forecast last Monday. One TV station said that Tuesday would be 'partly sunny' while the other station said, 'partly cloudy.' On Tuesday there was sunshine but also clouds." "I remember," Martin cut in, "It was partly cloudy." "No, it wasn't. It was partly sunny because part of the sky was sunny." "Which part?" Martin asked, sitting up in his chair. "The part between the clouds." "Gotcha! You just said that part of the sky was cloudy so that would mean it was partly cloudy." "Somehow, Martin, I knew you might say that, so I got on the internet to the National Weather Service web page. They had a list of all the words used to describe the sky.

As it turns out 'partly cloudy' means the same thing as 'partly sunny.' It's sort of like your half glass of apple juice. Some people prefer to look at the brighter side of things so that's why I say 'partly sunny'. I noticed that even some of the TV weather forecasters use 'partly sunny.'" Martin nodded his head. "But why use 'partly sunny' when you can only say that during the day? 'Partly cloudy' works in the day and night." "I didn't make the rules, I just looked them up. 'Partly cloudy' is the same as 'partly sunny.'" "Don't be so touchy. What else did you find on your mission?" "Some of the words that you hear in forecasts to describe the sky mean different things depending on the kind of clouds, and some phrases have a wide range of interpretation." "So does the weather." "...And that's why they are used," Miranda said. "Most of the other sky words are a little more clear. No pun intended."

"'Sunny' or 'clear' mean you'll hardly find a cloud in the sky. 'Mostly sunny' or 'mostly clear' means up to 70% of the sky is free of clouds. 'Partly sunny' and 'partly cloudy' tell you that 30% to 70% of the sky is filled with clouds." "I get it," Martin interjected. "That would mean then that 'mostly cloudy' says the majority of the sky is covered with clouds and 'cloudy' means that all of the sky is covered." "That's it." Miranda applauded. "Now that I've cleared the cloudy words, are you mostly done with that half a glass?"

Stratus clouds are those that are stretched out and stratified. They are wider than they are tall, and they tell you that the main motion in the atmosphere is horizontal. Horizontal motion is called advection. Stratus clouds don't get the love that cumulus clouds get. Cumulus clouds have an emoji! Stratus clouds are just as important, though. In their thickest form, stratus clouds are an opaque sheet creating an overcast sky blocking sunlight and the moon. This can keep a city cooler in the day. At night, thick stratus clouds act as an insulating blanket to keep temperatures from falling as far as they otherwise would. Stratus clouds are made of water droplets, but in cold areas and high altitudes, they are also composed of ice crystals. When they produce precipitation, it is usually steady rain or steady snow that continues for a while. Thick stratus clouds that create precipitation are called nimbostratus. Nimbo means precipitating. Nimbostratus can be middle-level or low-level.

We find stratus clouds at all altitudes in the troposphere. The highest ones are called cirrostratus. These are made of ice crystals and are often thin, and sometimes barely noticeable. Cirrostratus clouds do not produce active weather. They may dim the sun, but they usually don't totally block it out. The ice crystals in a veil of cirrostratus clouds can bend light in refraction. This creates colorful rings and arcs.

Delicate fragments or wisps of high clouds made of ice crystals are simply called cirrus. They are often feathery and don't cover all of the sky. In history, cirrus clouds have been nicknamed mares' tails, as in the tails of horses, and they've been called emissary clouds when they arrive in advance of a storm system. Cirrus clouds also frequently form from airplane vapor or condensation trails, as the ice crystals spread out and persist.

Middle level stratus clouds are altostratus. These may be thick or thin and they may also be a mix of water droplets and ice crystals. Thick altostratus clouds blanket the sky to hide the moon or the sun.

As air rises smoothly over mountains and large hills and then sinks, we find clouds that show the multitude of motion of the air that we normally don't notice. These clouds are called lenticular, which means lens-shaped. Lenticular clouds can stay in place, or they may form downwind from mountains and drift slowly away. When the air rises and sinks smoothly, the clouds persist in the rising portion as the wind blows through them. When the sun is low in the sky, lenticular clouds appear much more dramatic. Lenticular clouds are a type of wave cloud, simply named because the air behaves as waves of water do, with a broad up and down motion.

A special type of stratus cloud sits on the ground or on water and that's called fog. Fog is a cloud. That's all it is.

See? I could stop writing right now, but then there'd be a big blank white space below, sort of like a cloud. Hmmm. That might be effective.

In another analogy to music, fog would be like a tenor or baritone—the lower range. The only thing that makes fog different from clouds above our heads is that fog is at our feet. A person traveling up a mountain into stratus clouds would call the clouds fog, by definition. Dense fog is a hazard to travel by vehicle, ship, train, or airplane. Reduced visibility gives drivers, captains, and pilots less time to react to obstacles or oncoming traffic. When it's in your face, slow your pace, leave extra space. In fog, colors are muted, and objects blur together. That's why drivers must have headlights and taillights on—to be seen better. Pedestrians and cyclists are at more risk of being struck, and need to be vigilant when traveling in dense fog.

Fog is a type of mist. Mist is tiny water droplets suspended or blown in the air but not falling to the ground. Mist will dampen objects that it touches. What differentiates fog from plain mist is fog creates low visibility, not much more than 1/2 mile and sometimes as low as only a few feet. Fog can form in place when winds die down, heat radiates from the ground to space, and relative humidity rises to 100%. This is called radiation fog and it may be thin, patchy, or thick, in layers. This is common where the ground is moist or where cool air drains into low spots like valleys. Radiation fog lingers until the air and ground warm up by sunlight, or until wind mixes the air enough to lower the humidity.

Fog on a calm pond or lake, common early in the morning, is nicknamed steam fog. Even though it resembles steam, it's not. It's more of a mist. It's not unusual to find vortex tubes or spinning tubes of fog dancing along the water surface, like miniature tornadoes. Fog that sits over water and is blown inland by wind is called sea fog. This can be very thick. It's also known as advection fog since it travels sideways along the ground or water. It's disruptive to shipping and boat traffic. When advection fog slides inland over cities it becomes a problem for drivers and pedestrians.

The formation of fog and clouds requires tiny particles to give the water vapor a foundation to build on. These particles are often dust, pollen, sea salt, smoke, exhaust, aerosols, and pollution. They act as a nucleus, to make it easy for vapor to start to condense, so they are known as condensation nuclei.

Given the nebulous nature of clouds, it is normal to find hybrid clouds or layers and clumps of clouds that merge and evolve over time. Stratocumulus clouds, for example, are both tall, bubbly and stretched out, as a hybrid of cumulus clouds and stratus clouds. Stratocumulus clouds can darken the sky and then produce precipitation.

Undulating motion in air may produce clouds that line up in neat, even rows called cloud streets. The technical name for those is undulatus, as in undulation. The rows and bands could be compact and close together or they could cover wide areas of many miles. Cloud streets show you the air is rising where the clouds are, and sinking in between.

Turbulence is seen in clouds as ragged masses or fragments that may be random or appear chaotic. At times, the clouds show swirls of spinning air while other times they appear to splash back and forth. Turbulence is obvious in clouds like mammatus clouds. Mamma means breast-like. Mammatus clouds form on the higher overhanging portions or anvils of cumulonimbus clouds. They show moist air is sinking into dry air to form little cells. Instead of bubbling upward, mammatus clouds bubble downward. Mammatus can look odd and dramatic, depending on how light hits them.

Also dramatic are wall clouds. They are a lowered, smaller portion of cloud beneath powerful cumulonimbus clouds, commonly known as supercell thunderstorms. There is a lot of rising motion in a wall cloud so it's not unusual to see clouds ascending rapidly within it, if you can see anything. Wall clouds can be low or hidden by rain or trees or buildings. It's a good idea to hurry and scurry to get inside. Wall clouds can produce funnel clouds which may further grow into tornadoes.

Sometimes a cloud fragment resembling funnel clouds appears, but it doesn't rotate like a funnel cloud, and it doesn't need a wall cloud to form. It's nicknamed scud, and it's simply a fragment of cloud hanging beneath a thunderstorm cloud.

One of the more unique clouds is manmade. It did not exist a century ago. It is a form of cirrus cloud because it is far above the ground and made of ice crystals. Behind high-flying aircraft, the vapor from combustion condenses to instantly form vapor trails or condensation trails. They are called contrails, for short. This is an unintentional, yet normal, process.

Pencil thin streaks, high in the sky. Streamers of cloud from aircraft that are high. Criss-crossing lines and curves that resemble tails, this moisture condensed is known as condensation trails. Regardless of the number of engines on a high-flying aircraft, a contrail forms in the wakes of the wings so a contrail actually starts as two tubes of air. The longer condensation trails last, the more the ice crystals may begin to spread, and fall a little. Wind shear makes contrails spread apart and morph into unique and pretty shapes and a plethora of patterns. These vapor trails often merge with natural clouds, making it difficult to discern the two. The motion of the trail behind the aircraft tells us which way and how fast the wind is blowing at flight level. A cross wind will quickly push the condensation trail away from the path that the airplane is flying in. Contrails that are new and crisp can cast shadows on layers of clouds beneath them.

Even when air is dry near the ground, vapor trails are found above, worldwide, in any season, especially in high-traffic flight corridors. On the horizon, a new contrail from an approaching airplane may appear as though it is rising upward like a rocket trail, but that's just an illusion based on angles and perspective. How long the vapor cloud lasts depends on how moist the air is at flight level, typically well-above 25,000 feet. If air is really dry, there is no contrail. If air is moderately humid, you may see a short or medium contrail, stretching out a few miles. When air is very humid, condensation trails can stretch out dozens or hundreds of miles. The longer they last, the more they spread into unique and sometimes odd shapes. You can see ripples, sheets, rows, curves, and wisps, all of cirrus clouds that started as vapor trails. Those can lead to bright areas and colorful arcs in the clouds, due to bending of light in the ice crystals. Because they are many miles above the ground, approaching sunrise, vapor trails will light up first, and then they will darken last, after sunset, often with yellow, orange, or red tints.

We see more contrails worldwide than ever before because air traffic is greater than it ever was. There are many studies underway to figure out how the extra clouds created by airplanes might impact Earth's temperature balance, as they block a little of the incoming and outgoing radiation.

For a quarter century, there has been a conspiracy theory circulating that condensation trails are an attempt to harm or control humanity by releasing chemicals in the air as "chemtrails." The belief takes a little bit of science and research history and mixes it with a lot of myths and conjecture, often presented as facts. There has never been any evidence or logic behind this. Condensation trails are easily explained by science. Combustion releases carbon dioxide and water vapor. That's why you see water drip from the tailpipe of the vehicle in front of you. Water released into cold air causes the relative humidity to rise to 100% and form a cloud.

When an airplane flies for a while at the same level as a supercooled altocumulus cloud, you may find a dissipation trail. Dissipation trails may also result in lines of cirrus clouds that would stand out next to a sheet of altocumulus clouds. The nickname for dissipation trails is distrails.

While clouds in the troposphere show endless combinations of form, height, and texture, it's the view from space that underscores the artistic and atmospheric variety and patterns. The classic comma shape of a low-pressure storm system is found every day in multiple regions of both hemispheres. Near-symmetric clouds of a hurricane create a signature known well in the tropics. Smooth, long, curved bands of clouds trace the jet stream and steering winds aloft. From any perspective, clouds integrate hydrology and meteorology to announce the state of the atmosphere. Clouds display the dynamics of wind shear and the buoyancy of air. They make obvious the ups and downs and sideways motions of our atmosphere. Clouds are an integral component of the water cycle as equal parts science and beauty. Pilots and astronauts can safely say, "I've looked at clouds from both sides now…" Clouds are weather, and weather creates clouds, but there are many times when there are no clouds. After a bunch of cloudy days, you might wonder, "Where did the clouds go?"

For those of you who ponder the paths and lives of clouds, the answers depend on the weather pattern. Wind blows the clouds somewhere else, especially after a cold front passes through. Clouds also disappear by simply dissipating right where they are. The moisture that makes clouds remains in the air, but the relative humidity falls below 100%. That's typical when dry air mixes in or when high pressure squeezes down on the air and makes the temperature rise a little. There are unique cases where clouds form or disappear, but the ingredients are always above us.

Have you ever heard of a hole punch cloud? That's the nickname given to a layer of clouds that develops a circular dissipation hole. The technical name is cavum, meaning cavity. These odd clouds can be caused by aircraft passing through altocumulus clouds that are made of liquid water droplets, even though the air temperature is far below freezing. That's called supercooled. Supercooled water in clouds is not unusual, and the passage of a plane in supercooled clouds can lead to instant formation of ice crystals that slowly descend as fall streaks. The ice crystals eat away the water droplet cloud layer to form a circle.

The hole story

December 11, 2003. It's a crisp fall day in southern Alabama. Clear sapphire skies and light wind create a carpet of dew by sunrise. I wait until mid-morning to tackle my task of cutting down a small tree in my front yard. The day is perfect. Temperatures top 50 degrees by eleven am, and my wife is not home to protest the demise of the tree. It really was dead anyway.

Knowing the fickle nature of falling trees, I secure a guide rope, and the aid of my neighbor, Bob. I figure that if Bob could take down a full-sized tree that was ten feet from his house, my tree would be no problem.

As we sling the rope up and around the treetop, I glance south toward the Gulf of Mexico. Cirrus, cirrocumulus, and altocumulus clouds are appearing in the southern sky, along with a distinct contrail. This pattern is common in the winter when there's surface high pressure centered to our north, and a westerly upper air flow along the Gulf Coast. With added vapor and condensation nuclei from a few high-flying jets, contrails seem to appear as frequently as natural clouds.

I crank the gas chainsaw. The wail of the engine summons my neighbor Ed to lend a hand. There's something about chainsaws and men ...

The sawdust flies, mimicking a mini snow squall. It settles around the trunk like a dusting of flurries. With the usual notching, cutting, and tugging, the tree responds; first a crackle of splintering wood, then the thud of thunder. The tree meets the lawn. Thanks, Bob. Thanks, Ed.

Now it's noon. I look up again to the south. Perfectly framed between two stands of tall trees is a circle in the clouds. A hole! I had seen a few of these dissipation holes before but never one so perfect or large. Without hesitation, I dash into the house and return with my SLR camera and video camera. I rush to get the video camera on the tripod and hit record. Once that's set, I attempt to take still photographs. One problem—my SLR battery is dead, and I don't have a spare.

Again, I run into the house and return now with a plastic point-and-shoot 35mm camera. I manage to capture a few good shots on color negative as I lament not having my other camera with a full suite of lenses.

For the next 20 minutes the hole grows and drifts eastward. I watch and surmise that a jet flew through the altocumulus deck at a steep angle, delivering condensation nuclei and water vapor.

These holes are not too unusual when supercooled water droplets form into ice crystals and then fall as virga. Ice crystals rob neighboring supercooled water droplets of vapor to further their own growth, working outward in a circle. The end result is a feathery cirrus precipitation in the middle of an altocumulus or cirrocumulus layer. Dissipation holes are nicknamed "hole punch clouds." Similar ones have been photographed around the world, but this one is unique for its size and symmetry.

The combination of the size of the hole, longevity, and the fact that it happened at lunchtime make it the top local news story of the day. I know. At the time, I was the chief meteorologist at the CBS television station in Mobile. By the time I arrive at work two hours later, there are numerous emailed digital pictures from our viewers. Public interest in the strange cloud is so great that we post a special page on our website with viewer photos of it.

Some people think it looks like an angel, while others joke about a similarity with a scene in the movie "Independence Day."

To confirm my suspicions of the cause of the cloud, I go to the University of Wisconsin at Madison website for high resolution images from NASA's AQUA satellite. I find a beautiful illustrative image that I share with my viewers. Taken around 12:30 p.m. local time, the picture clearly shows the patch of cirrus clouds centered in the hole over southern Mobile County. Not only are there several well-defined contrails to the west, but what is even more interesting is a long dissipation trail in southern Mississippi, along with other holes in the clouds there.

A dissipation trail is the cousin of a condensation trail. This type of clearing line is often caused from a jet flying in the cirrocumulus or altocumulus cloud layer. The vapor trail causes supercooled water droplets to freeze into ice crystals and fall as cirrus streaks, eating away the cloud deck in a straight line!

What I learned: A little hole can become a big deal. Look up, and make sure you have good batteries in your camera! ...and that's the hole story.

If you are a cloud lover, do an online search for "WMO International Cloud Atlas," and you will be amazed at the variety and beauty of clouds!

Alan Sealls

CHAPTER 4

Seasons and Weather Cycles

Why shouldn't you tell a joke while ice skating in winter?
The ice might crack up.

One year is 365 days of change. Even songwriter Benard Ighner noted that in his classic song, "Everything Must Change." The sky, clouds, humidity, temperature, and wind change each week, each day, each hour, and sometimes each minute. Over months, the changes and extremes of weather are constrained by the season. Knowing the season in advance is simple because seasons occur on a perfect cycle, defined by the tilt of the Earth's axis on the orbit plane around the sun. Contrary to misconception, the varying distance between Earth and the sun does not cause seasons. The definitions of seasons we use are astronomical, based on the positioning of planets, not the actual weather. Nothing meteorologically significant happens on the day a new season starts, and seasons don't tell you what weather you'll get on any given day. Seasons are useful for knowing what type of weather is likely or possible.

The farther north you are of the equator, the more dramatic the change in seasons can be, since northern latitudes cool so much more in winter than equatorial areas. In summer, the North Pole is tilted toward the sun. Daylight hours are plentiful for the northern hemisphere and the sun is high in the sky. Sun rays are more direct and therefore more concentrated and intense than they are in the winter. This combination warms the northern hemisphere. Many more locations can warm into the naughty nineties. The North Pole gets 24 hours of summer daylight, but because the sun angle in the Arctic is less direct than it is at the equator, the North Pole warms but does not get hot. The hottest places are nearer to the equator where sun rays are most direct and strong.

In summer, the steering winds and jet stream around the northern hemisphere weaken because there is less contrast in temperature between the equator and the Arctic. Those winds are driven by the temperature difference, making them thermal winds. Following weaker steering wind, large storm systems move slower in summer than they do in winter.

Summer starts in late June on the solstice. It's a roasty toasty season that can deliver heat waves, as the humidity gives you a hug. The oceans near the equator warm enough in summer to produce tropical storms and hurricanes. These immense low-pressure systems have a wide impact as they move through and outside of the tropics. Millions of people in dozens of nations are affected by a landfall with threats of flooding and extreme wind.

Extreme heat is a summer threat that can lead to cramps, heat exhaustion, or heat stroke. Heat stroke can kill. Excessive heat is made worse by high humidity. This is where heat index comes into play—that's how hot you really feel. High humidity means your perspiration evaporates more slowly, so your natural cooling system does not work as efficiently on the Gulf Coast as it would in the desert Southwest, where low humidity creates the infamous "dry heat." It's the high overall humidity, which is really the dew point temperature, which controls your comfort on a warm or hot day. The higher the dew point, the faster your glasses and smartphone fog up when you step outside. Moisture condenses on them, since they cool the air touching them to the dew point temperature. That condensation is dew. When dew point temperatures are in the middle and upper 70s, the air feels heavy and you can barely stroll down the sidewalk without breaking a sweat, or is that glistening?! Either way, the expression you hear for summer discomfort, "It ain't the heat, it's the humidity," really should be, "It is the heat and the dew point temperature."

On an annual basis, extreme heat is the number one weather killer in the United States according to NOAA. The impact of heat stress is greatest on people with health issues, and on the elderly. Intense sunshine on a vehicle with windows closed on a warm or hot day can quickly send the interior temperature upward an additional 20 or 30 degrees or more. This is a dangerous and sometimes deadly scenario for pets or children left alone in vehicles.

In summer, while contending with scorcher torture, you may hear of a heat dome. Maybe the first thing that came to mind when you heard heat dome was the TV show called "Under the Dome." It was science fiction about a mysterious invisible dome covering a small town, trapping people inside. That does sound a bit like a summer heatwave that sends the masses indoors to climate-controlled spaces. A heat dome is simply a contemporary phrase to describe what years ago was just called a heatwave or a stagnant ridge of high pressure. Where does the dome come into play? High pressure is a mass of air that has uniform properties of temperature and moisture. It is like a bubble of air, sitting over a region. Air has weight and the weight of the bubble of air flattens the bottom part to the ground. Since the bottom is flattened and the top remains rounded, if you look at weather charts at different altitudes or in 3 dimensions, high pressure resembles a dome.

Based on community development of roads, industrial complexes, and shopping areas, along with your neighborhood's home density, and amount of tree cover, you often deal with a much different summer afternoon temperature than you might hear in a weather forecast. There is certainly a dramatic difference in how hot something gets in direct sunlight vs. how hot the air temperature is. I did a quick temperature survey around my yard one summer afternoon in Mobile, when the air temperature was 99 degrees. Using an industrial infrared thermometer, I measured blacktop temperature at 135 degrees. My wood deck was 130 degrees, while concrete was 115 degrees. Even in the shade, the blacktop was over 100 degrees, while the other surfaces were closer to 90 degrees. This explains why your vehicle thermometer shows more heat than the standard or reported air temperature. It's registering heat rising from the road. It also makes it clear that the duration and intensity of sun, along with the color, density, and texture of a surface, control how hot a surface gets. Hot surfaces emit heat to influence how hot it feels around your home, day and night.

In the days before clocks, sundials or shadow length and direction were used to tell time. Time was relative to sunrise and sunset. Sun time is different from clock time. "Solar noon" is when the sun is highest and most intense, halfway between sunup and sundown. Things in the sun heat up fast at that hour, and that includes people! Solar noon, however, is not the hottest time of day. There's a lag between the sun reaching its peak, and the ground warming to the highest temperature. The hottest time of day is a couple of hours after solar noon because the ground keeps warming, even as the sun gradually lowers. You notice that same lag where pavement does not instantly cool after sunset. It takes many hours, through the night, for it to really lower in temperature.

What you wear further influences how warm you feel on a hot day. Dark colors absorb more sunlight to increase the heat. Heavy material with long sleeves and pants would certainly be uncomfortable in the hot summer sun. To really be cool, wear light-colored, light-weight clothing that reflects heat, and allows your perspiration to evaporate.

Why don't meteorologists give the temperature in direct sun for all the people who work outside in the afternoons? Because many of the factors above play into the temperature. Every different type of location outdoors will have a different temperature. That's why air temperature in the shade is a world standard for weather readings. It's critical since it dictates thresholds for planting, harvesting, and freezing of agriculture. It's also a standard because there is no sun at night, for most of the planet. In fact, to give a temperature based on sun we would have to know whether you are on asphalt, concrete, soil, sand, gravel, grass, a roof, and we would have to know what you are wearing! And when it gets cloudy for a minute, everything would change fast, except standard air temperature.

Intense summer sunlight causes faster tanning and raises the risk of sunburns and long-term skin damage and skin cancers. Another summer hazard occurs when high pressure sits over an urban area where pollution is created by vehicles and factories. Light wind does not mix the air much, so sunlight interacts with vehicle exhaust and other pollutants to create ozone near the ground. Even though ozone is a form of oxygen, high levels of ozone are an irritant for people with breathing difficulties. Conversely, ozone is a natural component of the stratosphere. Way up there, it benefits life by blocking harmful ultraviolet radiation from the sun.

Fall or autumn begins in late September on the equinox. With a Latin origin, equinox translates to equal night. Astronomically, the equinox is when every point on Earth has equal hours of night and day. On the equinox, the sun is directly overhead at the equator but it's only on the horizon at the Poles.

There's a myth that only on the equinox can you balance an egg or a broom. People try it and find that it works, but if you were to try it on any other day, you would see it always works. With patience, you can balance an egg or a broom on any day of the year! The problem with a myth like that is it does not specify what type of broom, or what type of egg, or what type of surface they are placed on, and there's no science that supports it.

Fall weather has colder air building in the Arctic, resulting in westerly winds gradually increasing in middle latitudes. Cold fronts become more common, depositing colder air masses farther and farther south. Many of us perch on the precipice of a chasm of cold. Nighttime temperatures tumble to put forth a freeze. We layer on clothing when our lips quiver and bodies shiver. A freeze is when the air temperature reaches 32 degrees Fahrenheit or lower. With colder nights in the 30s, even above freezing, some areas find frost where the surface temperature of the ground or vehicles actually does go below freezing. Autumn in many regions brings relief from high humidity and heat of summer, although tropical storms and hurricanes are still possible. Those colossal cyclones are driven by warm ocean waters which retain much of the heat of summer. They move moisture around the planet and redistribute energy from the tropics to the middle latitudes.

The fewer hours of daylight and lengthening of darkness, along with cooler air, triggers the annual change in color seen in many deciduous trees. The green color from regular production of chlorophyll disappears. As the green fades, it reveals magnificent golds, oranges, reds, and yellows. The intensity of leaf color depends on what type of tree is dominant in an area, how warm or cool the weather is, and how moist the soil is. Frost may have a little to do with leaf color but not all trees in all places experience frost, especially before the colors change. As leaves fall off trees in advance of the coldest season, many northern locations begin to see frost while others begin to see snow.

Winter is the coldest season, and it starts in late December when the North Pole is tilted the most away from the sun. The 24 hours of daily winter darkness at the North Pole allows temperatures to reach their lowest levels of the year. The contrast between a very cold Arctic and warm equator drives the strongest westerly steering winds, between about 5 and 8 miles above the ground. The jet stream is the core of these winds, and it reaches its greatest strength as it meanders farther south, creating large and powerful winter storms. Winds in the jet stream can exceed 200 mph.

It's winter that forms and locks the coldest air at high latitudes. Polar or Arctic air is bounded by strong winter winds known as the polar vortex. These winds encircle the Arctic, in the stratosphere. While the polar vortex may sound like a gigantic cold windstorm, it is simply a band of powerful winds far overhead that meanders and changes shape. The polar vortex may allow Arctic air to slide southward to middle latitudes, often creating record cold, but not necessarily with strong wind near the ground.

Some of the largest winter storms form in the Southeastern U.S. and move along the Eastern Seaboard.

Guided by the jet stream, nor'easters draw warmth and moisture on the southeast side from the Gulf of Mexico and Atlantic Ocean, and cold air on the northwest side from the Arctic. As these winter storms travel northeastward, high winds along the coast typically blow from the northeast. That's where these storms get their name. A nor'easter can deliver high wind, intense snow, rain and lightning on one side, and bitter wind chill and blizzard conditions on the other. Large nor'easters and other intense winter storms may even have an eye, like a hurricane.

Winter storms produce hazards over a wide area that are not always related to heavy snow, high wind, or extreme cold. Freezing rain creates a glaze on surfaces that are at or below 32 degrees. That's an ice storm. A coating of ice or frozen sleet makes travel treacherous by foot or by vehicle. Layers of ice on a vehicle prevent practical or safe driving. Bridges and overpasses are at high risk of ice since they are not insulated by ground beneath. Their temperatures fall fast. Any wet, untreated road or highway can form a layer of ice when the temperature is below freezing for a long enough time. This leads to vehicle accidents from loss of control. Glaze stops trucks without tire chains. Even with tire chains, there's no guarantee that a truck can negotiate ice and snow, especially on inclines. An isolated accident can block major travel routes to strand motorists in the cold.

Even without storms, extreme cold can lead to hypothermia and frostbite. Hypothermia means your body temperature is below a safe level. Frostbite is when your skin literally begins to freeze. Dress in layers and stay dry. Exposed skin loses heat at a rapid rate, especially when the wind blows and creates an additional wind chill. The wind chill temperature you see or hear in a weather forecast is calculated only for humans — warm-blooded creatures, with moist skin. It's how cold the air feels to you, as your body heat rapidly escapes. The wind chill temperature does not apply to objects like houses, cars, or water. Those things will certainly cool faster with wind, but they cannot cool below the air temperature. Pets and outdoor and wild animals experience a different wind chill because many have fur coats.

Getting out of winter relieves the stress of cold weather problems. Spring is the season of rebirth and rapid growth in our environment for plant, animal, and insect life. The warming springtime air and soil, combined typically with more rainfall, gives a start to new plants and a restart to dormant plants and trees. Insects seek abundant pollen, but allergy sufferers don't look forward to increased pollen levels. In some regions, pollen leaves a coating on everything. It may also travel hundreds of miles in the wind before settling out of the air.

Springtime on the calendar starts in late March on the vernal equinox. Warm air begins to migrate northward, following the more direct rays of the sun. The spring season forms powerful storm systems that thrive on the contrast of warm and moist air heading north, clashing with cold and dry air heading south. The jet stream guides, supports, and agitates storm systems, so spring is when we see our highest number of tornadoes around the United States. Spring tornadoes are commonly fast-moving and more likely to occur in clusters or in large families known as outbreaks. Most occur in open areas but those that strike densely populated cities or communities with vulnerable homes like mobile homes, can be destructive and deadly.

Planting time

Any hint of spring weather makes us anxious to shed clothing and get outside. For many of us, we're excited to get our fingers in the soil, sow some seeds, beautify our yards, and grow something good to eat. Spring may not be as welcome for allergy sufferers, as pollen makes it the sneezin' season, following the freezin' season. As in other seasons, spring weather is an average, not a definite. A shift from a southerly wind over several days to a northerly wind for several days flips temperature.

"When is it safe to plant my garden?" That's the question from those who fear the freeze. I'm only a sometimes gardener but I know the answer relates to what you are trying to grow. Every plant has a different tolerance for a freeze or long periods of cold. You have to check the seedling freeze tolerance of whatever you plan to raise, as it relates to the average growing season where you live. The farther north of the equator you live, or the higher the elevation in mountains you are, the later your threat of freezes lasts in the spring.

Past averages and records don't dictate future conditions. The weekly weather patterns control when we freeze, but we can't reliably look more than 10 days ahead for a trend, let alone an exact low temperature forecast.

For some tender and young plants, you also have to account for the damage that frost can do. Frost depends on air temperature, humidity, and wind, along with things like tree cover. Frost may form when the nighttime temperature stays in the middle or upper 30s. So, when is a good time to get the garden going? What's your level of risk and how much effort are you willing to put into covering or moving young plants? Have patience to avoid risking wasted effort in starting an outdoor garden. Even with the regional rules-of-thumb, you still need to keep up with the daily forecast.

Cycles

Everything in weather repeats on a cycle. Some of the cycles are very consistent and easy to predict. Other cycles are thrown off by changes in the local atmosphere, based on weather systems. Seasonal cycles are more reliable. In the big picture, the cycles of cold in the winter and warm in the summer always occur. Hurricanes, in the summer or fall, happen every year. Snow in the winter is expected. Tornadoes in the spring are typical.

Daily weather cycles are driven by sunrise and sunset. Once the sun rises, it heats the ground which in turn heats the air. Temperature rises through the middle of the afternoon. Relative humidity falls as temperature rises. Bubbles of air ascend, and the clouds appear. By mid-afternoon, the weather is warmer and breezier with lower relative humidity near the ground. After the sun sets, the temperature falls throughout the entire night. Relative humidity rises. Winds die down. Clouds dissipate, and the coolest time of the night is at sunrise.

When low pressure systems are nearby, or air masses are moving, these cycles are shifted. If you keep a record of daily weather where you are, you'll begin to see regular patterns or cycles of movement and lifespan of weather systems.

In daily weather cycles, the coldest time of day is actually a few minutes after sunrise! Strange, but true. When people ask me, "How cold will it get tonight?" I often give the answer with a late night or early morning reference, or I just call it the morning low, since the lowest temperature is reached after the night is done. The next time you lay awake, pondering the universe, and unable to drift off to sleep, just get up and check the temperature every hour. Somehow, I don't expect you to do that, but at any time, you can pull up hourly temperature data online to verify the daily temperature cycle.

How sunrise is the coldest time of a typical day is simple. After sunset, the heat that the land absorbed during the day radiates out to space. With light wind, and low humidity, the air temperature will keep on falling until something heats it up. That something is the sun's rays. Why don't we heat up the minute you see the sun? Air does not absorb and retain heat very well, compared to ground. While air is partially heated by sunlight, most of the heat we feel is from sun rays, or radiation, striking the ground, so that the ground warms, and then the ground transfers the heat to the air.

Think of all of this as a pot of water on the stove. The water won't warm until you turn on the stove. If you take it to a boil and then turn off the heat, the water will continue boiling for a little while before cooling, until the heat is turned back on.

Atmospheric cycles depend on the region and the season. The summer season along coastlines has a daily sea breeze that pushes air inland, often forming cumulus clouds and afternoon showers. At night, the breeze reverses, pushing the showers offshore. In valleys surrounded by mountains, the cycle may include cold air draining down the mountains at night to form fog in the valley. The fog dissipates in the morning as the valley warms. Along the west coast of the United States, the cold Pacific Ocean waters contribute to low clouds that move inland at night and back out to sea in the afternoon. This humid mass of air is known as a marine layer.

Mathematically, any cycle is a series of curves known as a sine wave. If you take two things with the same cycle or frequency that each cause the same type of weather, you'll find that they interact just like sound waves. When the two are synchronized, their impact is increased to enhance the change in weather. But if the two are perfectly offset, they will cancel each other and not have much of an impact on weather.

In reality, when one of those cycles has a different frequency from the other, the net result is much different and may not appear as anything predictable. We see it daily in computer projections of wind around the hemisphere. This principle works out on every beach. Wind and bathymetry create multiple waves of different heights and wavelengths, or frequencies. As these waves interact, the outcome is far from uniform waves in a perfect rhythm. That is the perfect analogy for cycles in the atmosphere, but weather also has changes that are not cyclical, and that further complicates things. These events could be once every few months, or years, or decades.

Within our annual cycles are some that are not as regular. For example, if the average westward wind in the Pacific Ocean weakens, it causes a shift in the locations of highs and lows, and warmest water and heaviest rain in the Pacific. When the water temperature off the west coast of South America warms to levels above average, this is called El Niño because the warming off the coast is first noticed around Christmastime. El Niño translates to The Child or the Christ Child. An El Niño also shifts weather patterns around North America and much of the globe. Some places get wetter while others end up drier and warmer. Atlantic hurricanes and tropical storms become less frequent. For every region that gets beneficial weather, there's a region that gets undesired or stormier weather.

La Niña is the opposite of El Niño, but it too will shift storm tracks and high-pressure areas around the world. La Niña is when the central and eastern Pacific Ocean along the equator is cooler than average, with slightly stronger east to west winds. Those winds push the warmer water to the western Pacific. When this pattern persists over many months, it's known as La Niña.

El Niño and La Niña are influencers. Neither is a storm, nor a single day of weather, but they do shift average weather patterns. You can't blame or credit either for any single weather system. Neither is bad nor good by itself. The impact depends on how strong they are, what region you are in, and what season it is. Either could last a few months or many months. They don't happen on regular cycles, and they don't necessarily alternate.

Other oceans and continents experience shifts in weather from similar changes around the Earth. Once the shift begins, it's easier to predict the outcome, but actually predicting El Niño or other similar phenomena before they start is difficult. Predicting the end to them is hard too.

Cycles and their offsets to the atmosphere offer an endless combination of weather scenarios that we see throughout our seasons. A reliable cycle of seasons, driven by Earth's orbit and angle with respect to the sun, sets the boundaries of possible weather. In the southern hemisphere, the seasons are offset by 6 months from the northern hemisphere's. This is because when the North Pole is tilted toward the sun the South Pole is tilted away. When the North Pole is tilted away from the sun, the South Pole is tilted toward it. That is another of the great balances of Earth.

Within global, regional, and local atmospheric cycles, small differences in weather components create very different types of weather. That's what makes things interesting but that's also what keeps the atmosphere balanced to support life on Earth. Math and physics together make the changes predictable in a broad sense. The accuracy and precision of the predictions, however, is limited by our ability to measure and understand the relationship between pressure, temperature, humidity, and wind, and how they interact with land, water, and heat from the sun.

Alan Sealls

CHAPTER 5

Radar, Satellite, and Instruments

What should you do when the barometer is falling?
Catch it.

You are a mobile weather station. The hat that flew off your head gave wind speed and direction. The goosebumps or sweat on your forearm tells the temperature. Your hairstyle displays the humidity. When the humidity is high, it can blow your 'fro, unglue your 'do, loosen your locks, or relieve your weave. The sight of lightning followed by the sound of a rumble provide distance of a thunderstorm. The wet shirt—that's rain. The human body is well-equipped to sense the elements. We see, hear, and feel changes in the atmosphere. Hot, cold, wet, dry, windy or calm—all of these are obvious, but they are also relative to what season and what climate you are in. They are also relative to your physiology, and for that reason our bodies are not the most objective measuring devices. Do you feel a chill? Is the air humid where you are? Is it windy? These questions posed to a group of people are sure to return a variety of answers and probably spark debate.

A meteorologist has to have an unbiased answer to those queries, based on a uniform method of measuring the elements. Meteorologists worldwide use weather instruments to gather measurements and data and communicate them in standard formats. We use calibrated instruments and tools to provide objective and mathematical values to weather elements surrounding us, and those that are so distant that our senses can't detect them.

Radar is the tool used by meteorologists for locating precipitation. The word radar is an acronym for RAdio Detection And Ranging. It's a device that uses radio waves to detect the quantity of precipitation and determine the range, as in direction and distance. A traditional weather radar is a rotating dish, resembling a large satellite receiver mounted on a pedestal, which points outward just above the horizon. The radar dish is protected from high wind and hail by a radar dome, also called a radome. The radome is coated with a special paint to minimize interference with the radio signal. The radar emits pulses of electromagnetic radiation, simply known as a radio signal. When a transmitted radio signal hits a target, it is absorbed and then re-emitted by the target in all directions. The amount of the signal that returns to the radar receiver tells how big the target is. The time it takes for the signal to return is used to calculate the distance. What returns to the radar is also nicknamed an echo because the radar, in a sense, is listening for the signal, just as a bat emits a frequency and listens for it to come back.

All radars operate on the same principle. The radio wavelength of a weather radar signal is chosen to detect small particles like rain, snow, or hail, rather than large objects like airplanes. The more precipitation found in air, the larger the area displayed on a radar screen. Colors on the radar show how intense the precipitation is. Typically, blue and green are lighter precipitation, yellow and orange are moderate, red is heavy, and purple is intense.

Weather radar displays are seen on TV, internet, and on smartphones. Sometimes what you see is not always precipitation because radars also detect flocks of birds or bats, swarms of insects, as well as aircraft. They can even see smoke particle plumes from large fires. At night or in calm weather with very high humidity, the radar beam refracts or bends downward toward the ground to give a return pattern called ground clutter. That's when radar detects trees, buildings, and even the ground, and makes it appear as precipitation that is not moving.

Powerful radars, like those used by the U.S. National Weather Service, can send and receive signals up to 250 miles away. All radars are limited in distance by the curvature of the Earth. With distance, the radar beam gets higher above the ground, so at some point the radar cannot "see" storms and precipitation on the horizon. The radar becomes less precise, too, with distance because the beam spreads out vertically and horizontally. Think of it as the way a flashlight beam spreads out with distance. A large radar beam samples a larger area so distant details cannot be resolved. For this reason, a network of radars is used in many countries, like the United States, to get full coverage of precipitation. Weather radars are literally blocked by mountains that rise in their beam path, making it impossible to detect what is happening on the other side.

Doppler radar can determine the horizontal motion of the precipitation. Its common use is to determine if there is rotation within a cumulonimbus cloud that may eventually lead to a tornado. When a tornado is big or near a radar, the radar gives a signature of rotation that meteorologists can recognize using the Doppler portion of the radar. Wind-blown rain that is moving fast toward or away from the radar will send back signals at different frequencies. It's the same principle used in police radar to estimate the motion of a vehicle. When there is rain spinning fast in a circulation, the portion moving toward the radar sends back a higher frequency signal, while the portion moving away from the radar sends back a lower frequency signal. This is known as the Doppler Shift, named after an Austrian physicist in the 1800s. A Doppler radar display shows the shift as adjacent but different bright colors in a very small area. This is called a couplet.

The newest type of radar adds a feature called dual polarization or "dual pol" for short. A dual polarization radar separates the signal emitted into horizontal and vertical components and that allows it to determine the shape and type of precipitation better. For example, the dual pol radar can discern large raindrops from hail, or snowflakes from rain. However, what the precipitation is by the time it reaches the ground is controlled by the temperature on the way down.

Radars can display patterns that suggest a tornado might be happening. Only in some cases can radar detect spinning debris in the air. That debris signature verifies a tornado. The most certain confirmation of a tornado comes from a human being who sees it.

Satellite

Day and night, a constellation of weather satellites watches Earth. Many countries have their own satellites and wisely share the data for the benefit of the world. The satellite most used on weather broadcasts is a geostationary type. These orbit the planet at the same rate of Earth's rotation so that they stay over the exact same spot, with the same view. In the United States, they are called Geostationary Operational Environmental Satellites—GOES. From a distance of over 22,000 miles away, these satellites send cloud images back to Earth on a regular cycle. We see the entire hemisphere, medium views of the country, and closer views of active weather. Some of these images are routinely updated every 5 minutes.

A satellite cloud image makes it easy to see the evolution of weather systems, when looped in a series over time. It shows weather patterns over the oceans, long before they affect land. Satellite images present the complexity and beauty of clouds.

Another type of weather satellite orbits on a different path at a much lower altitude. Polar orbiting satellites don't stay over one spot; they orbit north and south across the poles as Earth rotates beneath them. Polar orbiters create a swath of images that only covers a part of Earth, twice a day. What they lack in frequency, they make up for in resolution. At a typical altitude of 500 miles, these satellites display features not just within clouds but also on land. They easily show flooding, dust storms, large smoke plumes from fires and volcanic ash, thick fog areas, ice on rivers, icebergs in the ocean, and snow cover on the ground.

From polar orbiting satellites, we find unique weather, like ship trails. Similar to airplane condensation trails, ship trails form when the exhaust from ship engines gives off smoke particles and water vapor in already very humid ocean air. A line of low clouds forms behind the ship and then often merges with natural clouds.

Polar orbiting satellites also give a detailed view of how wind interacts with small land masses. Downwind of mountainous islands we see the swirls of eddies in the clouds, just as you might find in a stream when water flows over and around rocks.

Both types of weather satellites also have multiple sensors for environmental elements like ozone, water vapor, and other gases. Some satellites carry sensors that point away from Earth to monitor solar activity. For a meteorologist, one of the most important sensors used by satellites is the one that measures the amount of heat radiated from Earth and from clouds in particular. The infrared sensor allows us to sense clouds at night. Since air temperature cools with altitude, clouds at different heights are different temperatures. Low clouds are warmer and high clouds are colder. The difference in temperature is displayed as shades of gray which looks much as it would for a regular picture taken of clouds using sunlight. This is how satellites give images of clouds in the darkness of night. Sometimes the satellite cloud image is enhanced with color for certain temperature ranges. This makes tracking cloud and storm features easier.

There are newer sensors, only on some satellites, which are able to capture images of clouds at night using light reflected from the moon. Other sensors can detect lightning flashes. Many satellites carry search and rescue transponders to aid in aviation and maritime incidents. With new satellite sensors and perspectives from space, meteorologists are constantly learning more about weather, but also finding new questions.

Weather balloons

To fully understand weather, a meteorologist has to know temperature, pressure, humidity, and wind, not just at the ground but also in the air, at multiple heights, around the globe. Weather sensors are launched by balloon twice a day at hundreds of locations on our planet. Through international cooperation, at zero Zulu and 12 Zulu, or Universal Standard Time, or Universal Time Coordinated, this upper air information is collected. The data from weather balloons is shared worldwide, as weather knows no political borders.

A small package containing calibrated weather sensors and a transmitter is attached to a large balloon filled with helium or hydrogen. After release, the balloon ascends. Wind carries it dozens and sometimes hundreds of miles away. Data is transmitted back to Earth in real-time to a radio receiver that tracks the balloon. This data results in a vertical profile of weather data. The instruments are called radiosondes and the newer ones use GPS.

A weather balloon easily reaches heights between 10 and 15 miles. That's much higher than commercial airplanes can fly. Some rise as high as 20 miles. At higher altitudes, the air pressure is so low that the balloon swells and eventually pops. A parachute carries the instrument package down to the ground at a safe speed. Many radiosondes land in forests, lakes and oceans and are never recovered. Those that are recovered in the U.S. can be mailed back to the National Weather Service to be refurbished and reused. Weather balloons are old technology, but they provide critical, accurate data.

Weather instruments

The most basic tool used by meteorologists is a thermometer. It's a meter that measures therms or heat. We call it temperature. You may have a thermometer with a liquid like mercury or alcohol inside a sealed glass tube. The liquid expands with heat and contracts as it cools. Other thermometers have a coil that is made of two different metals sandwiched together and connected to a pointer. This bimetallic design takes advantage of the principle that different metals expand and contract at different rates with temperature change, so when the air warms, the coil stretches out to push the pointer higher.

A hygrometer is used to measure humidity. It is a meter that measures hygro (water). Professional hygrometers traditionally have used animal hair or human hair connected with a mechanism to a pointer to give relative humidity. As water vapor increases in the air, hair lengthens. This is why on a very humid day it's harder to keep your hair in control. It lengthens, but not uniformly, too often ending up in curls! In dry air, the length of hair decreases. Hair, especially horsehair, has been used in hygrometers at least back to the 1800s. It may not sound very scientific, but hair is a fairly accurate tool in the true definition of "relative" humidity. Not many people like their locks to be a continuous experiment! That's where oils and conditioners come in to help give your hair control. They coat the hairs to interfere with the normal reaction of hair to humidity. For you that's a positive.

You may have noticed that some wooden doors or drawers in your home stick or creak when humidity rises. Any object that can absorb moisture will react to humidity changes. Basic hygrometers may also use an absorbent material bonded to a coil of metal. As humidity changes, the material changes length to stretch or squeeze the coil and make the pointer move to show a value.

Other weather instruments determine humidity by calculating the rate of evaporation. Once you know the humidity and the temperature, you can use a chart or formula to calculate dew point temperature. That's the temperature that air must be cooled down to in order to form dew. You can directly measure the dew point temperature by noting the exact temperature when dew forms, but that's not always practical. While dew point gives an indication of the absolute amount of moisture in air, relative humidity does not. Relative humidity indicates the amount of moisture in the air relative to the moisture itself, and relative to the air temperature. Think of a meeting room that has a maximum capacity set at 100 people. If you have 40 people in the room, then the relative capacity is 40%. However, if you divide the physical room in half, the maximum capacity drops to 50 people and the relative capacity increases to 80%, even though the total number of people did not change! Relative humidity rises when moisture is added to the air and/or when the air temperature lowers.

Wind has two components that have to be determined: direction and speed. Wind direction is fairly easy to observe using a wind vane. It's a simple instrument, often seen on older buildings and on boats, that points in the direction from which the wind is blowing. Some wind vanes have ornate and even whimsical designs with an arrow to show where the wind originates. Wind is named for the direction that it comes from, rather than where it is going. The wind vane used by a meteorologist may not be ornate, but it is typically connected to a dial or meter to record and display the wind direction.

Wind direction is also displayed at airports, using a windsock. It's a brightly colored tube of fabric connected to a pole, alongside runways. A windsock allows pilots to see the wind direction, to aid in safer takeoffs and landings. The amount of the windsock that stretches outward also gives pilots an idea of how strong the wind is.

Wind speed is measured by an anemometer. That is an instrument with spinning cups or propellers. The stronger the wind, the faster the spin. Many anemometers are built into a wind vane, resembling an airplane without wings. The wind vane points into the wind and the propellers measure the speed of the wind. This combination instrument is called an aerovane. Wind speed can also be determined by the frequency of sound when air passes over a tube in an instrument called a sonic anemometer. You've done that demonstration blowing over the top of a soda bottle.

Knowing the wind gust is as important as knowing the steady wind. That's added information on the strength of weather systems. Wind gusts can be more damaging than steady wind, especially depending on how frequent they are. Wind is air in motion and air has weight.

The instrument that measures the weight of air or air pressure is a barometer. That's a meter that measures air, using units of bars or millibars. Also measured in inches of mercury on a traditional standing vertical tube barometer, air pressure tells us how strong a high-pressure system is or how deep a low-pressure storm system is. Barometers, like those you would find in a home, use a sealed metal chamber with a partial vacuum. When air pressure increases, it squeezes the chamber. Springs and levers make the needle on the display move to a higher value. When air pressure decreases, the chamber expands and moves the display needle to a lower value.

Traditionally, these weather instruments were found at airports, since weather has remained critical to aviation. They would be housed in an instrument shelter—a northward-facing louvered box about chest high, over a grassy surface. The northward alignment was so that when the box was opened to read instruments, the sun would not shine directly on the instruments. Depending upon the needs of the user, weather instruments are now set up on instrument towers, buildings, factories, bridges, and roadsides. Many of the newer electronic or digital versions of the instruments are solar-powered and automated, housed in louvered, water-resistant compartments. With greater reliability and durability, and digital output, they have replaced the traditional instruments. This provides a great quantity of regular data, although the data lacks the visual sense and judgment of a human weather observer.

A rain gauge is another instrument and perhaps the simplest. It is a tube or container, left in the open, to catch rain which is then measured. Rain gauges that report data in real-time often use a miniature bucket that tips and empties each time it captures one one-hundredth of an inch of water. Every time it tips, it sends a signal to indicate the additional amount of rain that fell.

Other instruments have specialized purposes. A transmissometer is found at airports. It measures the transmission of light through air, simply called visibility. Visibility is reduced by particles. Common particles include water droplets in fog. Any solid or liquid or combination particle floating in air is called an aerosol. Aerosols can be natural like pollen or dust, or they can be particles like soot, ash, or pollution. These tiny objects in large number create haze.

Ceilometers are also found at airports. They measure ceiling of clear conditions for an airplane, which is the base of the clouds. Transmissometers and ceilometers are critical to safe aircraft operations.

More specialized weather instruments can measure ultraviolet radiation, ozone, and even the electric field in air. Lightning is preceded by an increasing electric field, so an instrument known as an electric field mill can give warning of the likelihood of cloud-to-ground lightning. There are also radio signal detecting instruments that detect actual lightning strikes to the ground.

Weather instruments are an extension of our senses. They have the benefit of being objective and consistent. However, all weather instruments must be calibrated, maintained, and properly positioned to avoid contamination of the data. For example, a basic home outdoor thermometer in direct sunlight will show a reading too high. A thermometer mounted on the outside of a wall will pick up some of the building's temperature, rather than just the air temperature. If a thermometer is placed too close to an air conditioning unit, there will certainly be false readings of air temperature.

A hygrometer placed adjacent to an exhaust vent won't always display correct humidity. A rain gauge that is under a tree will not accurately measure rainfall. Wind vanes or anemometers mounted too close to structures cannot measure the wind without a bias.

You may have noticed a naming convention in many weather instruments. They end with the word meter, and start with the word of the element they measure or the unit to measure it, like therm-o-meter, ceil-o-meter, bar-o-meter, hygro-meter, although the o is pronounced as ah.

All weather instruments allow meteorologists to create a record of weather and weather events. Instruments afford a universal method to measure, communicate, and share weather data around the planet. Weather instruments enable us to monitor thresholds for comfort and safety, and to define and catalog storms and weather events by magnitude. This data is used in studying past climate, and in making future predictions of climate and weather.

CHAPTER 6

Lightning and Thunderstorms

What kind of shorts would a thunderstorm buy?
Thunderwear.

Lightning gets our attention, although maybe not as much as the rumble of thunder that follows. Lightning is an artistic arc and spark of electricity that travels through air in unpredictable directions, at speeds beyond our comprehension. Every moment, somewhere on Earth, lightning is streaking across the sky in attempts to balance the electric field over the planet. Lightning occurs when positive and negative charges separate within thunderstorms to build a large difference in electric potential. The electric discharge of lightning takes on different shapes, paths, intensity, duration, and color. The strongest lightning strikes have hundreds of millions of volts of electricity.

The classic image of lightning shows it extending from a cloud to the ground, forking downward. Most lightning actually does not hit the ground. It flashes from one cloud to another as cloud-to-cloud lightning. It also travels from one part of a cloud to another part of the same cloud as intracloud lightning. Sometimes, lightning crawls horizontally, often underneath the anvils of thunderstorm clouds. The nickname for this is spider lightning. Lightning also flashes from a cloud to the air with the same purpose of moving electric charge. Like air and water, electricity is a part of our environment, but we don't notice it until there's too much and it suddenly moves.

Lightning is nature's warning to get inside. As deadly as it can be, it's fascinating from a distance. One of the many weather questions I get over and over is, "Does lightning go up or down?" It may be a phone call from a husband and wife who are debating, or an email from a college kid. The answer is yes, and yes. Lightning goes up. Lightning goes down. I see lightning all around.

The lightning we see reaching the ground starts as a series of weak electrical charge segments, stepping down from the cloud and branching outward at the same time, seeking a source of current. Your eye doesn't catch that because it is faint and happens in a tiny fraction of a second. What you do see is current shooting upward into the cloud. The upward flow of current illuminates something that was branching downward, so you process it as a flow of electricity downward. Think of the branching as tributaries of a river. Tributaries carry a substance into the main channel. This lightning process starts downward, but what you see is actually going upward.

Let's flip that. Literally. Lightning can initiate at the ground, branching upward. In that case, the current does flow downward to the ground. When I say ground, you can substitute soil, water, tree, tower, building, or person—most often the tallest object, regardless of what it is made of. Whenever lightning starts or ends at the ground, it's called cloud-to-ground lightning.

All lightning creates thunder, but when thunderstorms are distant or when the wind is blowing from you toward the storm, you don't always hear the thunder. If Simon and Garfunkel were meteorologists instead of musicians, "The Sound of Silence" may have been "The Sound of Thunder." *Hello lightning, my old friend. I've come to see you strike again. Because a vision brightly creeping, left its sounds while I was sleeping. And the sounds that were planted in my ears, still remain. Within the sound of thunder.*

In "Dreams," Stevie Nicks sings, "Thunder only happens when it's raining." That's poetic but it's a bad lesson. If I had written the tune, I would have penned, "Thunder only happens after lightning." I would have added another line, "Lightning can happen before the first raindrop, and lightning can happen in dust storms or in volcanic eruptions, where there is no rain." Maybe by then, the song title would have to be "Nightmares" instead of "Dreams." Thunder does happen when it's raining but it also happens before and after rain.

Thunder follows lightning. It's a boom, rumble, crackle, peal, or roll that can startle. Intense heat from lightning compresses air quickly to force an outward rush, in waves. Thunder is essentially a shockwave, like the sonic boom formed by a supersonic airplane that carries energy to vibrate air and your home.

It's not just one sound from each lightning channel. All portions of multiple lightning branches send out shockwaves of different strength and volume that overlap, interact, and arrive at your ears at different times, as well as at the same time. It's a stereophonic symphony driven by physics. The sound of thunder depends on the intensity of the current, the number of return strokes in the lightning channel, the length, shape and path of the channel, and the density of the air. The closest lightning strike creates an immediate bang after the flash. When a lightning strike is a single vertical nearby bolt, you hear an explosive boom that trails off. If a strike travels horizontally, where part of it is overhead but other parts of it are 5 miles away, you may hear rolls and rumbles that last 25 seconds.

Even in calm air, the sound of thunder travels relatively slowly, compared to the visible discharge.

You've heard that when you see lightning and then count, "One Mississippi, two Mississippi, three Mississippi ..." until you hear thunder, and then divide that number by 5, that tells you how many miles away the lightning is. Yes, that is true, since it takes sound roughly 5 seconds to travel a mile. The rule, however, is not a good way to forecast thunderstorm threats. It doesn't work well when lightning branches across the sky, or when there are multiple thunderstorms at the same time, or when you are monitoring a distant thunderstorm and a new one is forming overhead.

As children, we all probably heard mythical or spiritual descriptions of what thunder supposedly is. From posts on my Facebook page, Heather said, "My mother just told me thunder was the sound of angels moving furniture in heaven, and to stay inside so none would get dropped on me. I follow that sage advice today." Apparently, that has worked for Heather, since she's still around.

Cecilia said, "My cousin Kathleen's kindergarten teacher told the class that God was rolling watermelons under the bed." Charles heard that thunder was, "God rolling the watermelons out from under his bed." Which one is it, in or out? For Kimberly, it was described as, "barrels down the stairs." Richard's mother told him, "Thunder is the sound of God bowling." I suppose that would follow a strike.

Beware the myths and misnomers. Heat lightning is a misnomer. It's a common nickname for distant, often frequent, lightning that occurs on a warm summer evening, which formed by the heat of the day. The name makes it seem harmless. It's a myth that heat lightning does not create thunder. It's just too far away for you to hear it. The person near the thunderstorm hears thunder and is at risk of being struck by lightning.

Lightning that travels out of a cloud sideways, sometimes for miles, before turning to reach the ground, is known as a bolt from the blue, as it seems to come from a blue sky. These catch people off guard and also occur before and after rain, making them even more hazardous.

Cloud-to-cloud and intracloud lightning are sometimes called sheet lightning because the clouds diffuse the lightning channel to light up the clouds in broad sheets.

The lightning that is the biggest threat to people is cloud-to-ground lightning. The tremendous electrical current reaches temperatures hotter than the surface of the sun. When the nearly invisible flow of electric charge reaches down, branching in steps, nearing tall objects or the ground, streamers of opposite charge reach upward. One streamer rushes upward through the branching channel, completing the circuit and illuminating the entire branching structure to form the flash. Once the stroke is complete, more current may flow upward in rapid pulses which create the flickering we see in lightning. Each one of these is called a return stroke and there may be several to a dozen of them.

Lightning tends to strike taller objects since they provide a shorter path to available current. Metal does not attract lightning. That is a dangerous myth. It doesn't matter if the target is metal, wood, or rock. However, lightning that strikes metal can send current farther along the metal whether it is a fence, train track, or power line. In that regard, you don't want to be around large metal objects outside during a thunderstorm.

Lightning striking sand has enough heat to melt and then fuse the sand into a glass-like tube called a fulgurite. Fulgurites typically happen just beneath the surface of sand, and they may have the same branching structure that lightning displays in air. Fulgurites may be several feet to over a dozen feet long, but they are delicate because of their size. The diameter of a fulgurite is typically around an inch, but some are found that are more than two inches across. That gives you an idea of how narrow a bolt of lightning is. Some fulgurites are smooth on the outside while others are rough. The texture depends on the type of sand, and the heat generated by the lightning strike.

Trees are frequent lightning victims because they are tall. When lightning strikes a tree, it often vaporizes the sap, creating steam that literally blows the bark off. It may splinter limbs and even set trees on fire. Never stand under a tree for shelter in a lightning storm. Even near a tree can be dangerous. When lightning hits a tree, the current goes to the ground, where it can continue a path and electrify the soil.

Dig this lightning

You're checking out at the supermarket. The rhythmic beeping of the grocery scanner puts you in a trance. Your eyes meander toward the tabloids. Just beneath the miracle diet headline is one about a mystery trench in a cemetery. "Cable TV for the Afterlife?" This is not real, but the following is a true story. In my decades as a broadcast meteorologist, I've gotten a lot of viewer questions about weather phenomena. A call to my station, years ago, led to a fascinating investigation ending in mystery. A gentleman in west Mobile reported lightning had dug a trench in his family cemetery. I had heard of lightning trenches but never had I seen one.

Having been burned in the past by viewer errors and pranks, I went to the cemetery to see for myself. The furrow was easy to spot. It looked just like somebody used a trencher to sloppily lay a cable line. I was blown away by the length and depth. From the base of a tall oak tree, a gash in the ground ran straight, more than 60 feet, and branched outward. I was still a little skeptical, so I looked further.

Walking around the west side of the oak tree, I spotted where bark had been freshly peeled away in a vertical stripe along a high limb. It continued on to a larger middle limb. This is a common sign of a lightning strike, although midway to the ground the stripe disappeared. There was no burn or missing bark around the trunk, but at the base of the tree was a 12-inch-deep hole, about 8 inches across with a burn to the root. From this point, a trench ran to the east in a straight line, varying between 5 and 10 inches, both in depth and width. These numbers shrank as the trench started branching about 30 feet away.

Soil was randomly tossed out in clumps and chunks, as much as 8 feet. Roots along the path were charred. The main trench went to the corner of an in-ground crypt and traveled underneath the lid before exiting. Just short of the exit, it blew off a small piece of cement into multiple fragments, and then continued a few more feet to the base of a small oak tree next to a metal fence. Here it took a 90-degree turn to run along the fence, leaving another small trench, before disappearing several feet later. Adjacent to the fence, I saw a small hole under the metal leg of a bench. This may have been caused by smaller and more winding branches that left the main trench as much as 30 feet toward adjacent graves. As a group, the various paths mimicked the forking pattern seen in cloud-to-ground lightning bolts.

In retracing my steps, I was even more surprised to spot smaller portions of a trench on the opposite side of the tall oak tree, in the same straight line as the main gash, but apparently traveling in the other direction. Family members who stopped by the cemetery while I was there said they had seen a lot of lightning a few days earlier, with little rain. I asked if there were underground wires or pipes, which may have channeled the electricity, and they didn't think so. Lightning makes its own path, even without guides! One person did say there once was a house adjacent to the cemetery, but it had been torn down years ago.

Lightning trenches have been documented before and they are probably more common than not, but this one was remarkably straight and large. There very well could have been other smaller and deeper trenches with it, though. The deeper mystery is what the current did as it passed through the graves. Your guess is as good as mine, but I was not about to start digging in this cemetery!

Many of us have a personal lightning story. Mine is of probably the closest strike to me ever. I know this not just because the flash and the bang were nearly simultaneous, but because I later saw where it hit the ground.

It was just a regular morning of downpours and occasional thunderstorms moving through Mobile. I was awake in bed, contemplating another day at work, taking the quiet moment to

*listen to the rhythm of the falling rain, when all of a sudden …
POW! A blinding flash with explosive thunder. My house
shook. I leapt from bed, because I feared it had struck my house,
and immediately began inspecting. I have a friend whose home
was hit by lightning years ago when he was away. There was no
immediate sign, but a smoldering fire that the lightning caused
led to a total loss.*

*I searched and sniffed. No problems. I checked all appliances
and looked in the attic. I looked out the window and studied the
trees to see if any were hit. Nothing different there. I glanced at
neighbors' homes and trees to see if anything was awry. No
problems. The only thing I noticed was my cable went out. After
the rain and thunder stopped, I got out in the yard to see if
lightning had hit the cable line. No sign of that, as I traced it
toward the curb. One of my neighbors walking by said, "Come
take a look at this." "Uh-oh," I responded, knowing what I was
going to find.*

*The lightning had struck right in the middle of their
blacktop driveway, blowing out pancake-sized chunks of
blacktop. The neighbor said she was in the kitchen when the
brightest lightning and loudest thunder that she had ever
witnessed occurred, just 40 feet from her window. I asked if she
had felt anything, since people sometimes report a sensation of a
wave of static electricity. She didn't.*

*Right in the middle of the hole that the lightning created was
an orange cable line. Aha! That's why cable was out to many*

homes. The line was clearly visible and slightly frayed. It was frayed but I wasn't afraid. The storm was gone. The frayed cable suggested that current entered the wire and travelled to the nearby junction box. I followed the line to the side of the neighbor's house, to where the other utility boxes are, and saw no burn or exposed wires. That was good. Everything was grounded. It's good to be grounded in life, otherwise a scenario like that can be more damaging to things inside the home. A neighbor on the other side commented on social media that she did see a flash inside her house from the cable line. Wow!

No one was injured, and there was no other major damage, although a bunch of my hanging pictures were askew, and some shelf pictures had fallen from the vibration. All of this after a big boom boom in my room room, from regular Gulf Coast weather. Now you know why many older homes had lightning rods, what are now better-known as lightning protection systems.

Lightning safety

The safest place for a person during an electrical storm is inside a solid building. If lightning strikes near the building, the electricity may still travel inside through the phone wires, cable wires, gas line, plumbing, or electrical line. It may also hit a tree nearby and get under the building through the tree roots. If lightning hits the building directly, current may still get into any of the wiring or plumbing; so, indoors during an electrical storm avoid touching any appliance connected with wires to the wall. Postpone your shower and avoid using the sink and other plumbing. A nearby lightning strike can also induce current in wires and electronics, causing damage, without directly hitting the building. Induction current is what charges your smartphone without wires.

Lightning damage to buildings can be limited by installing a lightning protection system. A lightning protection system, LPS, is not just a few lightning rods. An LPS is typically a series of lightning rods spaced along the high points and corners of the roof. These metal rods are linked, often by copper or aluminum wiring that goes deep into the ground. This creates a frame or a cage around the building. A lightning protection system gives lightning a more direct path to the ground, should it strike the building. Metal does not attract lightning, but it is very efficient at conducting electricity. There are many different types of surge protectors used on homes and appliances, but they are designed for small spikes in current flowing through the wires, not the intense power of a direct lightning strike!

Outdoors, if you hear thunder then you are close enough to be struck by lightning. Most lightning injuries and fatalities occur before the rain falls and after it ends because that's when people don't expect lightning. Lightning frequently strikes outside of the rain and may travel miles ahead of a storm. The majority of lightning victims are young men simply because they don't go inside. They don't want to stop their recreation or outdoor work. They put themselves at risk unnecessarily. It's not just golfers, it's men playing baseball and basketball. It's roofers, boaters, swimmers, landscapers, and construction workers.

If you are caught outside in a thunderstorm and cannot get to a permanent building, a hard-topped metal vehicle is your best lightning shelter. It's a myth that the rubber tires protect you. You are protected by the metal shell of the vehicle which allows current to flow around it and then jump to the ground. This is the same reason why the passengers in a train, airplane, or ship are fairly safe from lightning. The enclosed metal shell of a vehicle acts as a Faraday cage, named after a British scientist in the 1800s.

Here are two other lightning facts, which can save your life. Lightning can strike the same place twice, during the same storm, or on different days. Rubber shoes don't protect you from lightning.

While most lightning victims do survive, burns to the skin can happen, either directly from the current, or from something metal the person had on them that heated up. People may be knocked unconscious. The worst scenario is cardiac arrest, which requires CPR. In the long-term, lightning survivors have to deal with the psychological aspect of having been struck. The strike can alter the brain on some levels, to where a person may have memory loss, or deal with a change in personality to some degree. A person may be changed but not in a way that others can see on the outside.

Lightning exists to balance electric charge in the atmosphere. It's a natural event that plays a part in the cycle of forest growth. Lightning causes woods and forest fires, which thin underbrush and allow new trees and plants to grow and thrive. However, when homes are in wooded areas, wildfires become a threat to people and property and then first responders. In any intense woods fire or wildfire, the combination of rising heat and water vapor, along with particles of smoke, contributes to increased clouds. At times, the rapidly rising air creates a cumulus cloud known as a pyrocumulus cloud. Pyrocumulus clouds are sometimes seen above active volcanoes as well. It's not unusual to find lightning in pyrocumulus clouds too.

Thunderstorms

Thunderstorms can be visually spectacular at a distance. In the daytime, they may appear as bright, crisp cumulonimbus clouds filling the sky. The tallest ones often have a flat top in the shape of an anvil. As thunderstorms approach, you may see a shelf cloud ahead of the dark base. Closer to them you feel the rush of cool and moist air known as the gust front. You can't miss noticing the electrical activity. The thunderstorm is nature's way of quickly redistributing heat and moisture, as well as electricity. Horizontal and vertical winds do the mixing and transport. Pilots avoid thunderstorms because of the rapid updrafts and downdrafts within them. Those can send an aircraft out of control. Even near a thunderstorm, air is turbulent.

Three ingredients lead to thunderstorms. Moisture is needed, as it is for any cloud. For a thunderstorm, you include an atmosphere that is unsettled or unstable. This instability often comes from either very warm and moist air in lower levels or very cold and dry air aloft. Then add something to lift or force the air to rise. This could be a cold front, dryline, sea breeze, outflow from a nearby storm, or a physical barrier like a mountain. It may also simply be strong heating of the ground from the sun. Given these ingredients, thunderstorms can happen in the winter, too, and produce bursts of snow known as thundersnow.

Hail from a thunderstorm tells you that the upper portion of the storm has temperatures below freezing. The falling rain and hail from high altitudes delivers much cooler air to the ground quickly. Thunderstorm clouds can grow anywhere from a few miles tall to 12 miles high. The taller the storm, the more rain it may drop, and the more energy it is capable of releasing in the form of lightning and wind.

There's no mistaking when a thunderstorm is overhead. The thick cloud blots out the sun, sometimes turning day to night. Winds increase and gust. The lightning flashes. The thunder crashes. In deepening puddles, heavy rain splashes. Wise people begin their dashes, for safe areas. It's unsettling for those without strong shelter. Add in the potential for hail and extreme wind and it's no wonder why thunderstorms are a hazard to aviation, transportation, and all outdoor activities.

A typical thunderstorm may produce winds gusting to 40 or 50 miles per hour. These will blow small or lightweight objects around or knock down small tree branches. When the wind gusts increase to over 57 mph, the thunderstorm is classified as Severe in the United States. That odd value is a simple math conversion from the criterion of 50 knots. Severe has a technical definition that does not often match peoples' perception of what severe is. The meteorological definition of a severe storm includes wind that is damaging—over 57mph. That's not to say heavy rain and intense lightning in a populated area is not news. From any strong thunderstorm, it would not be unusual to see small or medium-sized hail although hail doesn't always come along with high wind. Hail of 1" or larger is another criterion of a severe thunderstorm. Even that size hail can fall from severe thunderstorms without extreme wind. Wind, hail, and lightning are not always tied together.

Some thunderstorms are simply slow-movers with frequent lightning but little wind threat. Every time there is a thunderstorm over you, you'll get heavy rain and potentially deadly lightning. A single lightning bolt can knock out power. Heavy thunderstorms are not unusual, so meteorologists put the focus on the ones that go beyond obvious, become severe, and do widespread damage. To the average person, "severe" is any thunderstorm that is scary, loud, very wet and has a lot of lightning. While the meteorological term can be confusing, it is used to set thresholds for government action, and to classify storms. It is based on numbers rather than opinion, to make it objective.

Thunderstorms have personalities, like people. Some are disruptive and others may be annoying. Michael Jackson once sang, "Who's bad?" and when it comes to thunderstorms the question is, "Which is bad?" or, "Is any thunderstorm not bad?!" "Bad" could be damaging wind or a tornado, or a single stroke of lightning that hits your house or a transformer and knocks out power for hours. If you want cleaner air, along with fresh water for reservoirs, lawns, farm fields and lakes and estuaries, then thunderstorms, like most people, are more good than bad.

When many severe thunderstorms are likely in the United States, a Severe Thunderstorm Watch is issued. It typically lasts 6 to 8 hours and covers large portions of states or multiple states. The area could be as small as a few thousand square miles, or as large as tens of thousands of square miles. The watch gives notice of the potential threat, and it means watch the sky. It does not guarantee that a Severe Thunderstorm will form where you are. In the overwhelming majority of a watch area, few locations get the intense impact, but it's impossible to know in advance which those will be. Nonetheless, have a safety plan and safe area ready.

Once a severe thunderstorm forms, a Severe Thunderstorm Warning is issued. That covers a smaller area such as a portion of a county or a portion of a few counties in an area called a polygon. Warnings typically last from 30 to 45 minutes when there is a high or imminent threat of damaging wind, large hail, and occasionally a tornado. These polygons cover dozens or a few hundred square miles, but it is a much smaller area within that feels the impact of the threat.

Watches and warnings for severe or hazardous weather are seen on TV and on smartphones with weather apps. They are also broadcast on NOAA weather radio.

Small hail can coat the ground and turn it white and cold but large hail damages crops, wipes out farm fields, and damages vehicles and rooftops. Big hail of 3 to 4 inches signifies a stronger thunderstorm with intense rising and sinking air.

Severe thunderstorms can produce brief tornadoes. Even without a tornado, severe thunderstorm winds can gust to 70, 80, 90 mph, or more. This type of wind knocks down large tree limbs and occasionally trees with shallow or weak roots. Winds like these will also lift larger objects and throw them in the air. Mobile homes are at risk of being hit or crushed by debris, and at risk of being blown over in severe thunderstorms, if they are not anchored well.

In the summer, thunderstorms often develop in place, slowly drift at 10 to 20 mph, and then fade nearby. They don't move fast because the steering winds are light. At the end of their life cycle, we feel mostly cool air rushing downward along with rain.

Springtime thunderstorms usually travel faster, with the steering wind. These storms can move at 40 to 50 mph and as fast as 60 mph. From fast storms like these, damaging wind quickly becomes more likely. All it takes is a medium wind gust, added to the forward storm speed, to create winds approaching 100 mph.

Any thunderstorm can create gusting winds but sometimes those bursts of wind are extreme. When air rushes out all at once from beneath a thunderstorm, it is called a downburst. This happens when dry air enters a thunderstorm and then rain falls into the dry air. The rain evaporates, cools, and causes the air to become dense and heavy. The air plummets to the ground and spreads out in a straight line, in all directions. This burst of wind is called a straight-line wind. Straight-line wind can have the same sound, power, and damaging impact of a tornado. If the downburst occurs only in a small area, then it is called a microburst. Microbursts cover an area less than 2.5 miles across, and they last under 5 minutes. They are not easy to predict, and the strongest ones, with peak wind over 100 mph, can leave damage greater than that of a small tornado.

Downbursts that are not damaging are frequent. If a thunderstorm is moving 30 mph, and has a 30-mph downdraft, add those together to get wind in front of the storm in a straight line at 60 mph. Behind the storm, subtract the downburst from the storm motion to get air that is almost still. A downburst, especially in arid climates like deserts, is responsible for dust storms or sandstorms. The wind travels away from the thunderstorms, carrying sand and dust. The particles reduce visibility and leave a coating on everything. The most intense dust storms can arrive as a brown wall before they choke out daylight. Dust storms may also result from strong steady or gusty wind over dry land.

All high wind gusts and downbursts are a huge hazard to aviation due to sudden wind shear. This is most dangerous for aircraft during takeoff and landing.

Ordinary or severe thunderstorms that move as a line create what is called a squall line. Squall lines have larger areas of impact than individual thunderstorms. When thunderstorms produce a continuous series of powerful straight-line wind bursts over the course of hours that is called a derecho. A derecho is made of thunderstorms, but it is much more than severe thunderstorms or a supercell or a squall line. A derecho is a large line or band of powerful or severe thunderstorms that creates a wind damage area more than 400 miles long and more than 60 miles wide. The thunderstorms produce downbursts and sometimes some tornadoes. The word derecho is Spanish and roughly translates to forward or straight ahead. The damaging winds of a derecho are mostly in the same direction.

Derechos move fast, sometimes faster than 50 mph, giving people little time to prepare. As the intense thunderstorms in a derecho sweep across many counties and states, they generate steady or widespread wind gusting between 58 mph and 75 mph. The strongest winds recorded from derechos have been over 125 mph. Derecho winds can be hurricane-force, or higher than the wind of many tornadoes. Along with the extreme wind threat, derechos may produce large hail and flooding rain. The worst derechos impact millions of people, leading to billions of dollars of losses, making life and recovery difficult for the survivors.

Derechos are most common in the central and midwestern United States from late spring through summer. Derechos do happen elsewhere and in other countries.

Thunderstorms that merge, interact, and feed each other while moving as a group that persists for many hours produce a region of low pressure called a mesoscale convective complex (MCC). These are common in the plains of the central United States at night in the warm season. They supply rainfall for agriculture, but at the potential expense of wind and hail damage. A mesoscale convective complex could be the size of multiple counties, or half the size of a state like Nebraska.

Individual thunderstorms can become so intense that they generate their own environment. The largest of them rotate slowly and maintain a strong updraft. These are called supercells. They are in such perfect balance that they can last longer than typical thunderstorms, as they produce intense and severe weather. Supercell thunderstorms are known for producing tornadoes. Most thunderstorms do not become supercell thunderstorms, and most supercell thunderstorms do not produce tornadoes. In a thunderstorm process that's not fully understood, it takes a perfect mix of rising air, wind shear, and rotating air to create tornadoes. Tornadoes may come out of the rear of a supercell thunderstorm from a low-hanging cloud feature called a wall cloud. To make storm tracking even more difficult, not all wall clouds produce tornadoes.

Any thunderstorm with potential for great harm is watched carefully on satellite and radar. There's an organization of volunteer weather spotters in the United States called Skywarn. Their mission is to observe threatening weather and relay reports back to the National Weather Service to help improve forecasts and warnings. When you are outdoors or traveling, keep updated to warnings for severe storms with a NOAA weather radio or a smartphone.

The one certainty of thunderstorms is they all produce the hazard of lightning, and the sound of thunder, even if you can't hear it. Respect them and understand them so that you don't fear them. When I was a kid, I slept really well during thunderstorms. Now, I don't. It's not because I'm worried about a tree falling onto my house, but it's because my job as a meteorologist makes me have to respond to disruptive weather, sometimes on my time off. That's no complaint but it relates to the fact that broadcasters serve a variety of people, and some have anxiety fueled by thunderstorms, while others have a true phobia of stormy weather. The debilitating or paralyzing fear of thunderstorms is called astraphobia.

For those with fear of thunderstorms, warm, humid regions present a perpetually worrisome environment. There are frequent thunderstorms and, in some locations, there are hurricanes—organized thunderstorms, taken to a higher level. Regular thunderstorms are loud and windy, and seemingly more so when they wake you in the silence of night. The rumble of thunder may be the worst part for many people, although the boom tells you that the part that can fry your electronics and electrocute you has already happened—that's lightning. High wind with swaying trees is not pleasant for anyone.

For people with a true storm phobia, I would guess that the fear arises from the thought that a thunderstorm will do harm to them. A small percentage of thunderstorms do become severe with winds over 57 mph, and/or large hail, and/or a tornado. Thunderstorms can create damage, injury, and death. However, if you take the average annual number of people in the U.S. killed by lightning, thunderstorm wind, tornadoes, floods, and even throw in hurricanes, it's around 300, according to National Weather Service data. That's a tiny fraction of the over 30,000 killed in car accidents, logged in National Highway Traffic Safety Administration data. In the overwhelming majority of extreme weather that we face, the overwhelming majority of us get through without injury or death, especially when we prepare.

Alan Sealls

CHAPTER 7

Tornadoes

What is a tornado's favorite game?
Twister.

Tornadoes manifest a mystery in meteorology, and the magnitude of energy in nature. A tornado, by definition, is a violently rotating column of air from a severe thunderstorm that touches the ground. On average, more than 1,300 tornadoes occur in the United States every year, but most do little harm or damage. The majority happen in open areas and don't last very long. Tornadoes that strike densely populated areas can be very destructive and deadly. Those are the ones that we all hear about. The nickname for a tornado is twister.

A typical tornado is smaller than a soccer field, and stays on the ground for several minutes, to travel just a few miles. Most tornadoes move from southwest to northeast, somewhere around 30 mph, but the motion and speed will depend on the winds that steer the thunderstorms. The slowest-moving twisters spin in place, while the fastest ones travel at over 60 mph.

A rotating cloud hanging from a thunderstorm that does not touch the ground is simply a funnel cloud. Funnel clouds often come from a lowered cloud structure beneath the severe thunderstorm called a wall cloud. The pressure inside a funnel or tornado is low enough to cause water vapor to condense and make the funnel visible as a cloud of moisture. It is possible to have a vortex of air spinning on the ground without a visible condensation funnel.

Over water, tornadoes are called waterspouts, but there are different types of waterspouts. A traditional waterspout is a weak cousin of a tornado. Rather than forming from a large and intense thunderstorm, most waterspouts form from small or medium cumulus clouds on sunny or quiet days. If they move onto land, they are then called tornadoes but usually don't last very long. Fair-weather waterspouts are common in some tropical regions, and they happen over large lakes outside the tropics. In contrast, powerful tornadoes can cross water or even form over water, but those are driven by a severe thunderstorm, with a strong circulation.

Scud clouds are ragged, dark, cloud fragments, often confused for funnel clouds. Scud clouds may show rising motion, but they don't rotate. Funnel clouds and tornadoes rotate.

Days in advance, forecasters can project the risk for widespread tornadoes or organized severe storms in outlooks. Outlooks tell us that ingredients will likely be in place. They cannot be specific to what will happen because that will be controlled by how ingredients later come together, and if there are triggers to cause development. When the risk of multiple tornadoes is obvious in the U.S., a Tornado Watch is posted for portions of a state or several states. The watch means that everyone must watch for a threat and have a safety plan. In a watch, don't worry, be wary. These watches last from 6 to 8 hours, covering thousands to tens of thousands of square miles. In patterns where large, violent, and persistent tornadoes are likely that may cause loss of life, the watch is called PDS. That means Particularly Dangerous Situation.

A Tornado Warning is issued when a tornado is occurring or very likely to occur. Tornado Warnings typically last 30 to 45 minutes, and cover dozens to hundreds of square miles. Even when tornadoes touch down, and travel on the ground for many miles, the total amount of area that they impact is very small, far smaller than a Tornado Warning, which accounts for the uncertainty in where a tornado may move, dissipate and then reform, and impact. Some tornadoes do occur without being detected, and without a warning.

More tornadoes form in the United States than in any other country. The topography of the contiguous states allows moisture from the Gulf of Mexico to flow northward, east of the Rockies. Westerly winds crossing the high terrain of the western states dry out. They end up creating a broad wind shear of dry air above moist air, but that alone does not produce twisters. Large numbers, or outbreaks, of tornadoes result when low-level winds rapidly take warm, moist air north while a powerful jet stream creates a high shear of wind in both direction and speed, as you rise in altitude. This lifts the moist air to form thunderstorms that may slowly rotate. If a cold or even dry air mass moves into the warm air, beneath the strong jet stream, further rising motion is created to form numerous supercell thunderstorms. These supercells likely pull in some of the natural spin created by the wind shear to eventually go on to produce tornadoes. Researchers are still studying this. Unknowns remain.

The worst tornado outbreaks in history have produced more than a hundred tornadoes in 24 hours, over many states. Other outbreaks have been smaller but taken hundreds of lives. Some tornado outbreaks have done hundreds of millions of dollars of damage. Fortunately, outbreaks are not regular or frequent. Isolated tornadoes are more common, forming from strong or supercell thunderstorms with high wind shear. The possibility of isolated tornadoes is harder to pinpoint than the possibility of widespread tornadoes.

The tornado formation process may be as simple as wind shear creating invisible tubes of rotating air. Strong heating of the ground, or the interaction of outflow wind from thunderstorms, along with air rising into a thunderstorm, may lift the tube of spinning air so that it is vertical. As the tube is stretched by a growing cumulonimbus cloud, it spins faster, and the counterclockwise spinning portion becomes dominant. While the dynamics are not fully known, a lowered cloud base, known as a wall cloud, can develop in the column of spinning air, and from the wall cloud, a funnel cloud may stretch down to create a tornado.

On average, the highest density of tornadoes is in the central and southern Plains. Tornado tracks since the 1950s show why this general area of the country is nicknamed tornado alley. Tornadoes occur there regularly. There is no one definition for what a tornado alley is, so other tornado alleys exist around the country wherever relatively large numbers of tornadoes happen.

Exact wind speed of tornadoes is unknown as they occur. Regular radars are far apart and only scan above the ground to determine the storm circulation strength. The only time surface wind can be directly measured in a tornado is when there is a portable radar or an array of research probes nearby. This does not routinely happen, so a tornado cannot be ranked until it is done, and then the damage is analyzed. Tornadoes are ranked by how much damage they do, using a scale named after Japanese researcher Dr. Ted Fujita. The Enhanced Fujita Scale goes from EF-0 to EF-5. The greater the damage, the higher the rank on the scale.

Why don't we know a tornado's wind speed or rating while it's happening? That is because radar scans the severe thunderstorm that generates the tornado in the air, not at the ground. At best, for a tornado that is a mile away from a radar, radar would scan about 50 feet above the ground. For a twister at 10 miles distance from the radar, the scan would be around 600 feet up, and for one that is 50 miles away the scan would be around 4,000 feet high. This explains why radar cannot detect waterspouts from small cumulus clouds at the beaches. They adhere to the cliché of being "under the radar."

A high radar scan leads to less precision in tracking tornadoes that are more distant from a radar. Radar "sees" the engine in the clouds that drives a twister. In my standard shift car, I can, but shouldn't, get my engine to 5,000 rpm, but if I'm only in first gear, my speed won't be that high. On an interstate, in sixth gear, 5,000 rpm would produce a speed that would get me arrested! The engine rotation speed doesn't always tell the ground speed.

Unless you have a portable research radar and you can position yourself close enough to a tornado to point the radar toward the ground, you cannot know in real-time what the wind is. You have to wait for a damage assessment. Those assessments use damage indicators and extent of damage along with past research guides and tables listing how different materials and construction types respond to high wind. For example, a softwood tree snaps at a lower wind speed than a hardwood tree. At the same wind speed, a mobile home on blocks will be damaged more than a site-built home on a slab.

A shortcoming of rating tornadoes by damage is when a tornado moves over an area with few things that are impacted, it's hard to rank. That could be a huge parking lot with no vehicles, trees or light posts on it, a barren field, or a lake or bay. In those cases, often the best estimation comes from using radar winds at multiple levels, compared to past similar twisters, to estimate what the surface wind likely was.

Video can give preliminary clues as to the strength of a tornado, but even with that, you must know how things were constructed. We are all seeing scarier and dramatic crystal-clear video of tornadoes, from an increasing number of people chasing storms. Technology and data available on smartphones have made it a lot easier for someone with knowledge of the atmosphere to place themselves in an area where a tornado is likely. Your goal should be to use that same technology to put yourself somewhere where you are out of the path of a tornado!

The overwhelming majority of tornadoes in the United States are ranked as EF-0 and EF-1 which are considered weak, although their winds can exceed those of a minimal hurricane, up to 110 mph. While these twisters are comparatively small and typically brief, they cause significant damage to mobile homes. These tornadoes will also knock down large tree limbs and uproot small trees. EF-0 and EF-1 tornadoes account for more than three quarters of the annual tornadoes in the United States.

EF-2 and EF-3 tornadoes are strong enough to knock over trees, destroy mobile homes and create substantial damage to permanent homes, frequently damaging or destroying roofs. These are a greater threat to life. These make up about 1 in 5 tornadoes. They tend to be larger and sometimes more long-lasting than the weaker tornadoes. Creating debris, taking down trees and defoliating others, these and stronger tornadoes can leave a damage path visible by satellite, known as a tornado scar.

The most violent tornadoes are EF-4 and EF-5. These are infrequent, accounting for less than 2% of all tornadoes. While fewer than 10 happen in an average year, they take the majority of lives because of their intensity and size. Some can range between a half-mile and a mile wide. The longest-lasting tornadoes of this intensity travel dozens of miles. Well-built homes and buildings will suffer major damage, if not total destruction, from a direct hit. Trees can have the bark stripped off of them by the wind. Debris in the wind does further damage and injury. The best chance for survival from these violent twisters is in an underground or reinforced safe room shelter.

Mobile homes do not attract tornadoes but twisters that strike mobile homes or manufactured homes do much more damage than those that strike stronger site-built homes. Rates of injury and death are higher in mobile homes than in other homes, even for a tornado that has weak or moderate wind.

Large objects that are not anchored are easily moved by tornadoes. Glass breaks and shatters, creating deadly shards flying through the air. This is why you must stay away from windows. Wide-span roofs and large walls of unreinforced cinderblock in gyms and auditoriums can easily fail in a powerful tornado.

Some communities use emergency sirens to warn of tornadoes. While these are good for people outdoors near the sirens, they may not be audible indoors. In very strong wind blowing from a person toward the siren, the sound is not heard very well, if at all. Sirens may be struck by lightning and fail. Given the shortcomings of sirens, weather radios and smartphones with alerts are strong and timely tools for notification of tornado warnings. These are devices that can save lives.

In recent years, about 100 lives are lost annually to tornadoes in the United States. Depending on how active the season was, the number of fatalities has ranged from under 25 to as high as 550. Of the more than 1,300 tornadoes that happen, only a few dozen are responsible for fatalities. It's less than 2 percent. Even though population continues to grow, the average number of people killed has gone down due to better warnings, awareness, preparation, and better building construction. We do see more weather coverage of tornadoes on TV and on the internet than ever before. The proliferation of video cameras, doorbell cameras, web cameras, and security cameras allows us to see the actual violent impact of tornadoes. A tornado that results in loss of life is known as a killer tornado. It's those, and those that are very photogenic, that are widely reported on; however, the large majority of tornadoes do not kill, injure or cause large destruction.

Forecasting days and locations when tornadoes are possible and then where tornadoes are forming and moving has immensely improved in recent decades; however, the science behind individual tornadoes is not yet settled. Most rotating supercell thunderstorms do not produce twisters. A small number of tornadoes in the northern hemisphere spin clockwise rather than counterclockwise. Some tornadoes occur with lightning while others don't. This is why researchers conduct field studies to try to solve the puzzle.

For all we know and can sense about tornadoes, no technology and no meteorologist can accurately predict and detect 100% of tornadoes. Particularly for isolated or brief tornadoes that are far from radar, detection is not easy. Since the setup of the atmosphere is different each time there's a tornado threat, we don't always know the threshold of when the first tornado can form. There are certain conditions that make you sneeze, right? Predict exactly when you will next sneeze. You can't, but once you start sneezing, it's easier to predict if or when you will sneeze again. It is impossible by technology or human ability to predict every tornado, especially the first one, the same way that doctors cannot detect every disease. A desire to be warned for every single tornado requires many more tornado warnings to be issued, for anything that has the smallest potential of becoming a twister. You probably don't want that, just like you probably don't want your doctor to call you about every small thing that has a tiny likelihood of becoming a concern. When false alarms increase, people lose trust. It's a delicate balance.

There are storm spotters and storm chasers who document tornadoes and relay sightings to the National Weather Service. Chasing tornadoes has never been more popular but it is a very risky undertaking. Tornadoes and lightning are inherently dangerous. A careless or untrained storm chaser takes risks that can lead to injury and death, even through distracted driving or by being trapped on congested roadways with no escape route. Some storm chasers have lost their lives.

The key to surviving tornadoes is being aware of warnings and having a safety plan and location. As a society, we should build better and smarter to make buildings less vulnerable to damaging wind, and to give people more places to find shelter. The best tornado shelters in general are in the basement of strong buildings, away from windows. In homes without basements, the lower floor interior small rooms like closets, bathrooms, and underneath stairwells are generally safe locations. In high rise buildings, interior rooms and hallways are practical shelters. Remember the story of the 3 little pigs. The home that stood up was the brick home. Take that a step further. A home with poured concrete-reinforced walls gives the best protection against damaging wind, as well as greater protection from fire and termites. Within many new homes or buildings, a safe room can easily be constructed as a bunker. New home designs like open floor plan, and large bathrooms with lots of glass and large mirrors are taking away some of the safety found in older construction styles. In older bathrooms, you typically had an iron tub, and iron pipes in the walls. In older homes, closets were smaller which made them more of a safety cage than larger closets.

Tornadoes are intriguing. Fundamentally, they are spinning air with low pressure inside that causes a condensation funnel. Tornado prediction is getting better, but the capricious nature of twisters and their genesis leaves many mysteries to be solved. Meteorologists can see patterns that lead to days where tornadoes are likely, but to predict the formation and location of a single tornado in advance will take many more years of study and research. We may not ever know all the answers.

There's another type of funnel that may be mistaken for a tornado from a distance. It's known as a dust whirl or dust devil. The dust whirl is dust or soil suspended in spinning air contacting the ground. It happens most often on hot days when winds are calm. A rapidly rising hot air current, known as a thermal, draws air inward from along the ground. The incoming air takes on a spin as it approaches the thermal. The moving air speeds up as it spins around and closer to the center of the rotation. That's the same thing that happens if a skater is spinning with their arms out, and then they draw their arms in. They spin faster.

A dust devil is a concentrated region of low pressure, so the air not only spins, but it rises in the center. Most dust devils are weaker than tornadoes, but the stronger dust devils or dust whirls have wind over 60 mph that can damage rooftops and mobile homes. Some are even stronger. Dust devils can be as small as a patio or as large as a parking lot. The clue that they are not tornadoes is there is no thunderstorm above them. In fact, many dust devils happen on days where there may be no clouds. They are driven by heat rising from the ground, and may reach heights of 100 feet. Dust whirls are spotted frequently where soil is dry and loose, and wind is light. Rather than being made of water vapor they are made of soil, dust, or sand, so they take on brownish colors. Dust whirls are also called sand whirls, in deserts.

It's not unusual to see leaves and debris swirling in a parking lot or around buildings. These might look like mini-tornadoes or dust devils, but they are harmless. These are caused by swirling and fluctuating wind. The size and shape result from how air passes over and around buildings and nearby obstacles.

There's another type of whirl, which does have great danger. In large wildfires, intense flames pull in air rapidly to replace the column of rising hot air. The air approaches the fire and begins to spin to form a vortex of rising fire. This is called a fire whirl. Fire whirls increase the danger of any fire because they make the motion of the fire more erratic.

While tornadoes are a big threat in certain seasons and certain regions, the odds of seeing one or being directly hit by one in most locations are pretty low. Even while under a tornado warning, with extreme wind gusts and a loud roar, followed by damage at a single spot, that does not confirm a tornado happened. It's an assumption people make but the judgment of whether it was a tornado or not requires a review of radar, and a damage survey, over a wide area. Even just walking around and looking at damage is like trying to reconstruct a multi-vehicle collision. What seems to be is not always what actually was.

Most people will never experience the most violent tornadoes, but the strong and medium twisters can still be deadly and very destructive. The weakest tornadoes may still be disruptive and damaging on a smaller scale. The worst tornadoes are usually detected and tracked, with warnings going out to the public. In weather patterns likely to produce twisters, you must use TV, weather radio, internet, and smartphones to stay aware and informed. Have your safe area and action plan ready in advance. In regions that routinely or even occasionally get tornadoes, planning for them is critical, not only daily, but annually, and over decades when it comes to city planning and home construction.

Part of that long-term planning is assessing wind hazards from trees. Have your trees checked by a professional. Get recommendations for the best way to keep them healthy, and/or remove the trees that have limited life left. Some types of trees are notorious for falling or snapping in extreme weather. Healthy trees can be a defensive line to help shield your home from high wind damage but weakened trees will put you at risk.

Trees are a part of our lives that some people take for granted, until they are gone.

R.I.P. tree

Did you hear about Tree? He's gone. Just like that. It was only yesterday that I saw him. He was so vibrant and full of life. He had just budded and put out new leaves. I can't believe it. He was such an important part of our neighborhood. Standing stoical, some people took him for granted. All the kids loved him. He gave them shade in the summer. The birds sang songs to him. Squirrels played amongst his arms. We woke up this morning to the news that Tree was laid out, flat on the ground. Broken, never to rise again.

What happened, you wonder? It was a storm. A strong thunderstorm. Tree's friends all made it through, although they shed leaves in his memory. Why him, you ask? Was there a warning sign? Was there something we could have done to have helped Tree survive? Was there a severe thunderstorm or a tornado?

A closer view, after a morning of mourning, showed the root cause of Tree's demise. Tree was not as healthy as we had thought. We only judged him from his outside appearance. We had no clue what was going on internally. At the base of his trunk and along the main trunk, Tree was hollow. That, by itself, is not a reason why only Tree fell, but when combined with high wind, it's a very likely cause.

A few communities away, Tree lost entire clusters of distant relatives but those were all snapped, well-above the ground, and not uprooted. Where they snapped, the wood was solid. Ashes to ashes. Sawdust to sawdust.

A single tree that falls in high wind that shows rot, disease or a hollow trunk says the winds were not necessarily extreme, but the tree was in a weakened state. A solid tree with shallow roots, especially in saturated soil, will often be pushed over in high wind, to where the roots lift the soil and pop out of the ground. A healthy tree that is snapped tells us that the wind was really high, but one isolated snapped tree is not a sign of a tornado. If it's a large stand of snapped trees it could either be straight-line wind or tornado winds. The larger pattern of which direction snapped trees point tells us if it was a tornado, and that requires a survey over a wide area.

Storm chase

After having been on TV for over two decades, I finally had a chance to "storm chase" in June of 2011. I had wanted to do this for years, but my job never gave me the flexibility to head out to the Plains when the big storms were likely. For all my broadcast career, when there's bad weather I go to work. What is "storm chasing" for many people is for me "storm waiting" and "storm spotting." My goal was to put myself in a relatively safe position to see all the features of severe storms that are hard to view on the Gulf Coast, where we have lots of pine trees, low clouds, and low visibility. This was not about the thrill of chasing. For me, it was a photography expedition and a research mission. For everything I've learned in school, from books, and at conferences, the best way to really grasp something is to experience it—and I did!

I chose June since I had a weather conference in Oklahoma City, and because summer severe storms move a lot slower than those in the spring. Spring thunderstorms in the Plains can move at 50, 60 or even 70 mph. You really do have to chase those to keep up with them and that's not the safest thing to do. The storms I encountered were moving from 20 to 40 mph which gave me more time to observe and photograph, but also more time to plan an escape route if the motion or strength changed.

This was a solo undertaking. I invited my wife but she, like most people, could not travel the way I needed to travel. Day to day I didn't know where I would end up after the sun went down, so I didn't always get the best of motels. Going after storms is like fishing. It requires a lot of patience, time, and luck. Some days you get absolutely nothing. Storm chasing also requires that anyone who goes along has the same tolerance and respect for risk as the leader. Some of us live on the edge and that's not me.

When to get out of the path of a storm and which way to do so is not something that should be debated!

For a meteorologist, it's not too hard to figure out spots where storms should erupt. It's easier than ever with forecast discussions and real-time updates from the Storm Prediction Center and local National Weather Service offices, added to the fact that we can read these and see radar on smartphones. In fact, there are a lot of weather apps for phones that not only show you radar, satellite, and lightning, but they use GPS to show exactly where you are on the same map. This technology is incredible compared to what existed a decade earlier.

Caution: There are places in the open Plains where there is no cell signal and that's where it takes the skill of a meteorologist to use basic observations to ensure you are not putting yourself in danger. Storms are not fully predictable, and thinking technology will be available and 100% accurate is risky. The ever-present danger in storm chasing is lightning. Your eyes are critical to watch motion and development of clouds. Your ears give you the cue that distant lightning is producing thunder. Even your nose can clue you in to wind direction if you happen to be anywhere near a cattle, hog, or chicken farm! Your skin lets you know when humidity or temperature rises or falls.

My daily travel routine was to check the weather charts to see where, within driving distance, the biggest storms were likely to form. I drove an average of 400 miles each day for five days. My journey took me from Oklahoma to Kansas to Colorado to Kansas to Nebraska to Kansas and then back to Oklahoma. Most of the driving was on U.S. highways or on interstates. I was happy to find many of the state highways and even county roads allowed swift travel on smooth pavement. Often times my intentional detours put me on a gravel road but whenever I pulled off the main route, I checked my navigation system to make sure it was a public road and not a dead end. Gravel roads can become mush after the gush of rain, so I didn't let myself get caught in that situation. I never drove more than 90 minutes without stopping for a break to stretch and let my mind re-focus. Since the larger part of my mission was photography, this happened naturally.

The first thing that struck me on day one was how windy it was, even without a storm. You'll notice lots of wind turbine farms in Oklahoma and Kansas.

Day 1 was hot in Oklahoma. Most of the afternoon, the temperature was over 102 degrees. From Oklahoma City I made my way into Kansas and watched a dust storm near the community of Kismet. Storms never formed there so I travelled eastward toward Greensburg, Kansas, on US-54 and intercepted a severe storm near Mullinville. It was producing a lot of hail, so near the outflow winds the temperature had cooled to the middle 60s. Using my radar app to check the strength of the storm, I made a calculated decision to drive through the hail to get to the "quiet" backside of the supercell. The hail was tremendously loud, but it was no larger than a dime. Had it been much larger I would have had dents, and had it been really big I could have lost the windshield. On the west side of the storm, I awaited a rainbow as the sun sank low, but I missed out on anything decent. With the winds gusty and fluctuating on the backside of the storm, I felt the temperature rise and fall probably 20 degrees in less than 10 seconds. Amazing. Dodge City was my overnight stay.

Day 2 had me heading to Denver (or really Fort Collins) where a colleague runs a weather research lab. Going westward on US-50 to Lamar, Colorado, was a good drive since the sun was behind me. I tried to coordinate my routes so that I could limit the morning and afternoon sun in my eyes. From Lamar, I went north on US-287 to Kit Carson for lunch. Everything was calm and smooth until a roadwork delay near Hugo, Colorado, put me behind by 45 minutes. I was not in too much of a rush and that delay actually helped me catch a severe storm crossing I-70 east of Limon, Colorado. The leading edge had already crossed the interstate, but I could see the green tint typical of storms with a lot of hail. The severe storm was very photogenic on the leading edge. I exited to get in front of it but the roads dead-ended, so I could only watch it move away. I wasn't totally alone. One of the

Denver TV stations had a crew on the same road and there was also a chaser and a Skywarn spotter. We all tend to think alike. I got into Fort Collins near sunset and spent the evening with my colleague, Walt, an expert in lightning, and a fellow avid weather photographer.

Day 3 started peacefully. I slept late and took my time heading out of the Denver area to Kansas. It was a typical bright beautiful Colorado morning with low humidity. I marveled at the snow on the peaks of the Rockies. By late morning, storms had already started forming just south of Denver International Airport, so I took the time on a gravel road to get good photos of a crisp cauliflower-like cumulonimbus cloud in the distance to the east. That was the direction I was going, so on my journey back to Limon, I was following storms that produced blankets of hail as they moved eastward. I didn't want to drive into the storm, so I exited I-70 and did a zigzag, first south, then east, toward Aroya, Colorado. This paid off. Not only did I catch the aftermath of a hailstorm, but I was able to set up about a mile from a severe storm and watch a wall cloud form, followed by a brief rope funnel that never touched down. A storm spotter pulled up alongside me, since I was in a good area to see things, and we chatted. Heading east on US-40, I followed this storm and it redeveloped and caught me in a hailstorm. I managed to find shelter from the hail under the canopy of a closed gas station. There was one man already under there in a pickup truck. When I jumped out of my car to take pictures, he smiled, waved, and offered me his hard hat. Good sense of humor. I was safe from hail, but I did get soaked since the rain and hail were blowing sideways. I had dry clothes in the car, so I was able to change as I headed back to Goodland, Kansas, for the night. On the way, I did catch my rainbow somewhere near Bristol, Colorado.

Day 4 had the setup for more intense storms forming over northern Kansas and southern Nebraska. I left Garden City going north on US-83 without being sure where the best place would be to target. I did notice a feature, an old outflow boundary, on radar that was just north of a warm front. I figured once the warm front moved north and met it, by midday, storms would erupt. I was right. As I got into Oberlin, Kansas, I chuckled as I saw one of the armored "tornado intercept vehicles" racing westward along US-36. I figured it was heading to storms that had already formed north of Denver. I stayed on my course and wound up in McCook, Nebraska, around lunchtime. That's where I expected storms to form, but just like boiling water in a pot, it would take a while. I had lunch then went to Wal-Mart and browsed. I checked the updates on radar and satellite then went to the city park and relaxed in the shade for a couple of hours. By late afternoon, clouds started growing fast so I headed north on US-83 toward Maywood, Nebraska. I was driving parallel to the storms which had become severe, and got good pictures of rain shafts and wall clouds. After about 45 minutes I checked my phone and noted a Tornado Warning had been issued just a little to the southwest of where I was. It was the same storm I was photographing. I needed to get west and then south of the storm but there were no major roads that went west. I did see a lot of gravel roads but that meant trying to beat the storm to get around it on roads that might not be in the best of shape. Nope. I chose to go south and then west on the main highways, but a funny thing happened. I drove south and stopped a few times for photographs. The next thing I know, the entire National Severe Storms Laboratory storm chase convoy shows up behind me and stops to set up their equipment. I knew I was in the right spot. Sure enough, I begin to see the broad rotation in what is now a southeastward-moving mesocyclone (wall cloud). The motion of the storm was now bringing it toward me. It was fascinating and almost mesmerizing, but I kept a constant watch on my escape route on US-83. The storm cell was only moving about 20 mph, so I was able to study it. There appeared to be a second wall cloud just west of the one I was watching but neither produced a tornado that I could view. I did see brief spin-up dirt clouds from what could have been microbursts or gustnadoes. The parent thunderstorm crossed US-83 north of McCook as I listened to the tornado sirens near McCook. By this time, there were several other storm chasers and Skywarn storm spotters. Of the couple dozen vehicles on that stretch of road, most of us were there for the storm. The convoy and other chasers departed and went east behind the storm. I had had enough for the day, so I headed into McCook to spend the night. At the motel I checked into, the clerk asked if I wanted a lower floor. She said all the other guests were a little nervous after the Tornado Warnings and wanted ground-level rooms. I had a pretty good idea that the tornado threat was ending so I took the 2nd floor, with a view, of course. A few more severe storms passed north of McCook, so I went back out to get some lightning shots after dark. In watching the local TV stations later at the motel, I realized that very few people could figure out which counties were under watches or warnings based on the small maps in the screen corner since so many counties were displayed. In the Plains, so many counties are square and don't have unique shapes so it's even more difficult to pick out your county unless you really know geography. It reminded me that people who travel probably have no clue what county they are in and that's why cities make better landmarks for locating storms, than counties.

Day 5 was my final day. I awoke to a Severe Thunderstorm Watch. As I showered and thought about the safety rules we preach about taking shelter in a bathtub, I realized that the tub I was in was resin or fiberglass and not very durable. Once back on the road, my mission took me back toward Salina, Kansas. I travelled from McCook, Nebraska, to Oberlin, Kansas, to Norton, Kansas, through Hill City, Kansas to I-70 east. Once I got to Hays, Kansas, a severe storm had formed right over Hill City that prompted a Tornado Warning. It was a very photogenic supercell cumulonimbus. I debated going back but I decided to continue eastward to Salina where more storms were likely that afternoon. Sure enough, more storms formed on another boundary. These moved slowly northeastward toward Junction City as a cluster, but didn't show any signs of producing tornadoes. They did produce small hail. I was able to position myself ahead of one of them that moved through Abilene, Kansas. By the time it got there, I was under a fairly safe cover where I could get good pictures while staying dry. The end of that day produced a large display of mammatus clouds. Just at the point where the setting sun appeared briefly to produce color in the mammatus clouds, I pulled over to the side of the road to take pictures. I'm always cautious about snakes but that wasn't the problem. After walking 20 yards from the car in tall grass, I realized that what I thought were gnats were actually mosquitoes and they were vicious. That quickly ended my day.

Each day was a new adventure with sights and sounds and smells unique to the central Plains. I learned a lot. With few distractions, I enjoyed nature and got to be a tourist, passing through many small communities. I also got to watch many of my colleagues on TV who I see at annual weather conferences.

So, what about the pictures and video? There were a bunch that were artistic and educational. Many ended up in science magazines, newspapers, websites, and TV programs.

My advice and caution: Storm chasing is not a sport. It is not for the average person because of the danger of the weather, and the risk of accidents by distracted drivers. There are companies that offer storm chase tours and a few even guarantee that you'll see a tornado. If you decide to do it, make safety your priority. Have respect for nature, traffic laws, and private property. Keep in the back of your mind that you may be witness to tragedy and be forced into action to warn a community, help or rescue injured people, and maybe even save a life.

Alan Sealls

CHAPTER 8

Hurricanes

What did the hurricane say when passing over an island?
I've got my eye on you.

Hurricanes are colossal clumps of clouds, large spinning storms that form in the tropics of the world's oceans. They are found mainly in the warm season when the ocean surface temperature is at least 80 degrees, and wind shear above the ocean is light. Hurricanes carry huge masses of moisture as they redistribute energy. Their genesis is regions of low pressure known as tropical waves, which typically drift westward. Within the tropical wave, clusters of thunderstorms may persist for several days as a tropical disturbance. A tropical disturbance is little more than a region of unsettled weather. They are very common, but most tropical disturbances don't develop any further.

If the air pressure lowers in a tropical disturbance, the thunderstorms may begin to rotate as a group, developing a circulation around a common center. Weather forecasters focus on these and designate the stronger ones as "invest" — that's short for investigation, meaning it is

177

something that will be investigated and watched closely. Invests in the Atlantic and eastern Pacific are given numbers for tracking purposes. The numbers go from 90 to 99, and then start over again. These numbers are mainly for meteorologists worldwide to be consistent in identifying and modeling the invests.

Sometimes you'll hear a letter following the invest number, in something like Invest 92L or Invest 99E. The letter designates which ocean basin the invest is in. L is for Atlantic, E is for eastern Pacific, C is for central Pacific, and W is for western Pacific. This naming convention was never intended for the public, but the internet changed that! You see more and more technical things like this that might seem odd. From internet and media outlets, the use of outlooks for investigations and disturbances extends the amount of time you hear about something in the tropics before it develops, if it even does develop. That can psychologically wear you down, because the detail we want for landfall comes far later than the initial identification and early forecasts.

If the wind within the circulation of an invest is persistent, along with falling pressure, then it becomes a tropical depression and it is given a different number. Tropical depressions are numbered each hurricane season, starting with the number one. The number is used and communicated to the public to keep track of the various depressions that may be going on at the same time. The main threat from tropical depressions is flooding rain. Winds are designated to be no higher than 38 mph.

Storm names

When a tropical depression strengthens, and the steady winds exceed 38 mph, then it becomes a tropical storm and is given a name. Names help the public to understand that it is no longer a depression. The name also makes it easier to tell systems apart, as there may be multiple systems simultaneously. In the various ocean basins around the world, different names are used for each storm. Tropical storms are universally called tropical cyclones. Cyclone can also be used to describe any low-pressure storm system like a tornado or winter storm, but it's most commonly used for tropical depressions, tropical storms, and hurricanes.

Hurricanes and tropical storms were first consistently given names in the 1970s. There are some years where there are 3 or 4 active storms at the same time in the Atlantic Ocean basin, (which includes the Caribbean and Gulf of Mexico). Names definitely help us tell them apart. Isn't that why we name children?!

In the days of platform shoes and disco balls, most meteorologists were men, and they chose to name storms after women. I think the women's liberation movement put an end to that. Male names were included in storm name lists for the Atlantic, starting in the late 1970s, to create a format that continues today. Storm names are alphabetical. Every other name is male, every other name is female. Names starting with q, u, x, y, and z are not used, based on the relative few numbers of them.

When a tropical storm or hurricane is devastating in lives lost or damage dollar amount, the name is retired from the list, to prevent further angst if the storm name were to ever be used again. There will not be another storm named Frederic, Katrina, Dorian, Michael, Camille, or Allen. Otherwise, the list of Atlantic basin names is recycled every 6 years.

You do hear more tropical storm names that are not pronounced the way most Americans probably would have guessed. The reason why is storm names are chosen by contribution from meteorologists representing all of the countries impacted by tropical storms. Tropical weather in the Atlantic is international. Prospective storm names go to the World Meteorological Organization (WMO), a 70-year-old agency of the United Nations. The names are voted on, and so too is the decision to retire names at the end of a season. This is how we end up with names that are common in countries that speak Spanish, French, Portuguese, English, and their various dialects. The WMO issues a pronunciation guide to limit debate! Notice, no names are Italian, Mandarin, or Swedish. That's because Atlantic tropical storms don't impact Italy, China, or Sweden. Every ocean basin with tropical storms has a separate list of names.

In the western Pacific, where hurricanes are known as typhoons, you would hear names like Ewiniar, Fung-wong, and Krovanh. In the Indian Ocean, where hurricanes are called cyclones, you would hear names like Burevi, Kyarthit, and Thianyot. These, too, represent names found in the multiple nations of those regions. So, what's in a hurricane name? History, diversity, and respect, if it is headed toward you.

Storm winds

The steady or sustained wind in a tropical storm may range from 39 to 73 mph. One benefit of a tropical storm is the same as one of its threats: rain. A tropical storm that progresses inland provides fresh water hundreds of miles away from the coastline which helps agriculture and water supplies. The threat is when too much rain falls in any one spot. A slow-moving tropical storm or tropical depression can drop 2 feet of rain within a day. Within a tropical storm, the average wind may not be extreme but individual thunderstorms can create further damaging wind gusts. Tornadoes are possible, too, in the bands of rain that feed into the storm, known as feeder bands. The steady wind from tropical storms approaching land pushes water onshore and creates higher tides and coastal erosion.

Tropical depressions and tropical storms sometimes have features or an internal structure found in regular regions of low pressure, making them a hybrid type of storm. These are called subtropical depressions or subtropical storms, but their impact is otherwise the same if they strike your home.

When tropical storms increase wind speed to 74 mph, they become hurricanes. Hurricanes are known as cyclones, or tropical cyclones, or typhoons, in other regions. A satellite view frequently shows a clear area in the center of the storm. That's called an eye. It's created by subsiding air, warming, and lowering the relative humidity, along with the force of the spinning motion pushing air outward. Within the eye, there are often many smaller swirls of clouds. There's a balance between the bands of rising air in the thunderstorms feeding into the hurricane and the sinking air in the middle. While the strongest wind is typically in the wall of clouds around the eye, known as the eyewall, the size of the clouds on a satellite view does not always tell you the intensity of the storm. In addition, the eye of a tropical cyclone can be hidden by cirrus clouds from the ring of thunderstorms around the eyewall. Sensors on some low-orbit satellites use microwave signals to penetrate the clouds and measure wind speed and direction, for a more accurate assessment.

There's a myth that the northeast quadrant has the worst or strongest wind of a hurricane. That is correct only in some cases. What's more correct is the right-forward quadrant often has the strongest wind impact because the forward speed of the storm can add to the average wind speed.

In a typical hurricane moving over land, the wind increases with height as there are no obstacles or friction to slow it down. Very tall buildings experience higher wind at higher floors. Whenever that high wind is transported to the ground, in gusts, you get extreme winds, likely with a roar. The roar of any wind signals possible imminent damage. The longer you hear that sound, the longer that extreme wind lasts, and the worse the impact may be.

During hurricanes, some people recount hearing a roar like a freight train, or fighter jet. Many believe it to be a tornado. People who have gone through direct tornado strikes, not in a tropical storm or hurricane, sometimes describe having their ears pop (due to sudden pressure change), and hearing a roar like a train. Wind and gusts in a hurricane are equivalent in strength to that of a tornado and they do also cause sudden pressure changes. This makes it nearly impossible to hear a difference between a straight-line wind gust, and a tornado wind, when one is in the middle of a hurricane.

From a tornado, especially one moving quickly within a hurricane, you would expect any roar and direct impact to last just seconds, and certainly well-under a minute. If you are hearing a roar in a hurricane that lasts multiple minutes, perhaps intermittently, it won't be a passing tornado. It would be streaks or bands of straight wind, that on occasion can spin up brief vortices like tornadoes that still don't last that long in one spot. These are not something that often show up on radar, so they are very tricky to confirm. Any suspected damage from these or a tornado is often hard to separate from that done by other wind of a hurricane.

When hurricanes and tropical storms lose their tropical structure, they become extra-tropical or post-tropical but may remain powerful storm threats.

Saffir Simpson scale

Hurricanes are ranked by their wind only, using the Saffir-Simpson Hurricane Wind Scale. On the low end of the scale is Category 1. The most intense hurricanes are Category 5, and these are much less frequent. A Category 3, 4, or 5 hurricane is considered major. The relatively small increments in wind speed result in much larger increments of wind force and total damage. The category by itself does not account for the additional threats of intense rain, tornadoes in feeder bands, and storm surge.

The Saffir Simpson Hurricane Wind Scale has been used for over 40 years to give the category of a hurricane, based only on the highest sustained wind speed, anywhere within the storm. It does not account for wind gusts. It was named after Herb Saffir, an engineer, and Robert Simpson, a former director of the National Hurricane Center. The scale originally included storm surge, but storms like Hurricane Katrina made it clear that the landfall wind and storm surge don't always match up.

If you use the scale for what it is, just a wind scale, it's fine. If you want more information, it's not. There are ongoing research projects and proposals to improve or replace the scale. Hurricane threats include wind, storm surge, rain, and tornadoes. How does one put all of those into one scale, as the threats can all be independent? How would you rate the impact of these scenarios:

A weakening Category 1 makes landfall with high storm surge and produces dozens of tornadoes, hundreds of miles inland.

A Category 5 weakens to make landfall as a tropical storm, but stalls to leave 40 inches of rain.

A fast-moving, small Category 3 hits in late November, when trees have fewer leaves, and steadily moves inland, with no tornadoes and little rain.

It would be a tremendous undertaking to replace the Saffir Simpson scale with an impact scale because construction type and age, population density, land cover, tree health, age and type, soil moisture, time of year, bathymetry, and topography have to be accounted for. Those all then have to be integrated with precise projections of flood, surge, tornado, wind, storm radius and storm motion. Impact means also somehow accounting for human action in preparation for things that, if not completed, might result in projectiles and debris, clogged drainage, overturned trucks, blocked evacuation routes, and runaway boats and barges.

Worldwide, tropical weather systems are tracked using their positions based on latitude and longitude. Lines of latitude run east and west around the Earth. Latitude tells us how many degrees north or south of the equator a storm is. Lines of longitude (also called meridians) run north and south and meet at the poles. Longitude tells how far east or west of the prime meridian a system is.

Storm formation

Throughout hurricane season the typical formation zones of tropical storms shift but that doesn't mean a tropical storm can't form outside the zones. In any location, warm water and light wind shear play a role in helping these systems start, but large amounts of dust can temporarily weaken or prevent the growth of a tropical storm or hurricane. The key word is "can," which is not the same as "will."

Dust plumes from the Sahara Desert of Africa are periodically blown westward, into the Atlantic Ocean. Huge clouds of dust blow off the Sahara Desert, carried by the same winds that push hurricanes westward on a transatlantic journey. Those are known as the trade winds. It's routine for summertime, although newer high-resolution weather satellites now make obvious what has been happening for millions of years. The dust leaves Africa in a mass of warm, dry desert air. That combination disrupts and diminishes development of tropical systems that it nears. The dust, however, does not totally stop tropical storms in the entire Atlantic.

In the countries of West Africa, and just offshore, the dust is an immediate threat to health because it may blot out the sun as it fills the air, creating respiratory issues. The farther westward it travels, the less thick it becomes, but the health impacts extend into multiple nations. In the many times each year that dust is lifted from the ground to enter the Atlantic from North Africa, it totals hundreds of millions of tons, according to NASA. For reference, an aircraft carrier "only" weighs about 100,000 tons! How far can it travel? Very far. More than 5 thousand miles. Saharan dust frequently moves through the Caribbean to sometimes reach Central America, the Gulf of Mexico, the Bahamas, and any of the eastern United States. Moving at about the same speed as a tropical storm, it takes two to three weeks from the coast of Africa to reach the U.S. Gulf Coast, if it holds together. Along the way, minerals, carried with the dust, fertilize land and water. Too much of those minerals can cause harmful algal blooms, like the Red Tide known on the west coast of Florida.

Too much Saharan dust is a concern for people with asthma and other breathing issues. It may trigger allergies. Bacteria is also carried with the dust. Bacteria can be good, bad, or neither. That, along with exactly how dust influences tropical weather, is still being studied.

While satellites may show large, solid sheets of dust moving across the eastern Atlantic, when it does reach U.S. shores, thunderstorms mix up the air, to further dilute and then help rain the dust out of the sky, so you shouldn't expect it to be a solid, uniform thick cloud. Saharan dust creates milky skies with muted color, but it may magnify color toward red for the rising and setting sun. There are other particles and pollutants that lead to haze, so not every haze you see in summer is African dust, but some you've seen in the past certainly were.

The Gulf of Mexico loop current influences tropical systems in the eastern Gulf. That's a current of warm water that flows from the western Caribbean, through the

Yucatan Channel, extending northward or northwestward into the central or northern Gulf, before looping clockwise to head southeastward, through the Florida Straits, then up the eastern seaboard as the Gulf Stream. Sometimes the pool of warmer-than-average water within the loop current can break off and meander in the Gulf. When tropical storms or hurricanes move over the loop current, they commonly strengthen from the enhanced warm moisture. Hurricane Katrina was an example of that, but it was an extreme example.

An ocean or marine heat wave is like a heat wave over land, except it happens over water when stagnant high pressure and light wind and sunshine cause ocean temperature in a region to warm. It builds over days and may last for weeks. Tropical systems that move over the resultant warm water have more fuel to grow and persist.

The lifespan of a tropical storm or hurricane may range from a day to several weeks. For that reason, tropical storms or hurricanes may travel only a few hundred miles while others travel thousands of miles. For the United States, some of the strongest hurricanes have origins in the eastern Atlantic off the coast of Africa. If you trace them backward, you find that many of them were actually regions of thunderstorms in low-pressure moving across the continent of Africa before developing over the ocean.

In forecasting tropical storms and hurricanes, computer models generate projections that are helpful but never 100% accurate. There are many rules of thumb for predicting tropical weather. A warmer ocean tends to create more or stronger storms. Wind shear generally weakens or destroys tropical storms. A slow-moving storm is more difficult to predict because there is more time for more possible outcomes. A slow-moving tropical storm or hurricane at landfall creates more storm surge and more rain than a faster storm.

Rich in moisture, tropical depressions, tropical storms, and hurricanes may leave a foot or two of rain after landfall, not just at the coast but also inland. It doesn't take a strong storm to produce extreme flooding. It's the large and slow-moving storms that carry the highest risk of floods, where rainfall can be three to five feet, separate from storm surge.

Steady wind over the ocean creates storm surge near and on land. Storm surge is when the winds of a tropical storm or hurricane push water onshore and uphill, and into rivers and bays. Even without rainfall, storm surge can raise water levels by many feet. Depending on the shape of the coastline, the bathymetry (underwater topography), and the strength and duration of the wind, the highest storm surges can be over 30 feet with battering waves on top of that. When storm surge is added to the regular tides, it's called storm tide. High storm tide increases destruction from moving water. The low pressure of a storm increases storm tide by a small amount, but most of storm surge is caused by wind.

Storm surge damage can be staggering, where entire homes simply disappear. Some ground-level homes may float briefly while being pushed. Others can have the ground floor gutted by moving water with only the roof surviving. Moving water with battering waves is a destructive and deadly combination. Low bridges can be no match for battering waves. Bridge roadway spans that are normally not too high above a bay or river can be compromised by powerful waves on a storm tide that punches upward to gradually push them off their piles.

In New Orleans, a city partially below sea level, water spilled into the city when levees failed during Hurricane Katrina. It was a delayed outcome of storm surge, which created a deadly flood, lasting for days.

Given the risks of damage, injury and death from tropical storms and hurricanes, forecasting them is critical. The unique challenge of predicting tropical cyclones is that they form in parts of the world where there is sparse detailed weather data. There are some buoys in the ocean, but they are very far apart. Ships can measure weather and gather data, but they stay away from strong storms, for good reason. Satellites give us the most information when storms are out at sea. Using a technique of analogues, comparing the size, patterns, and cloud temperatures to past storms, a meteorologist can estimate the strength of a hurricane or tropical storm. Any actual direct measurement adds great value to the estimate. Without knowing exactly what a tropical storm or hurricane is doing when it's in the ocean, it's a huge challenge to forecast what it will do when it gets close to land.

Hurricane preparation

Hurricane season for the United States is from summer through fall. On a calendar, that's from June 1 to November 30, but there's no rule that says there can't be a tropical commotion in the ocean outside of those dates. These dates are when the combination of warm ocean water and light wind shear above the oceans most frequently allows tropical systems to form. The majority of tropical storms and hurricanes do not impact the United States, but they do threaten the many islands of the Caribbean and Bahamas, as well as Bermuda. On a map of all hurricane tracks in the last century, what stands out is that most of the storms follow similar curved paths and most of the storms stay in open waters of the Atlantic. It's also clear that hurricanes and tropical storms strike Mexico, parts of central and South America, and Eastern Canada.

Coastal residents need an emergency supply kit that includes food and water, important legal documents, photos and documentation of possessions, and cash. Those in storm surge and evacuation zones must know their evacuation route and have a plan. When evacuating, the idea is to travel outside the evacuation or danger zone. You don't have to travel far, just far enough to get away from the immediate impact and threat. As hurricanes and tropical storms move inland, they carry their hazards of flooding and tornadoes. Plan early with family and friends. If you wait too late to evacuate, you might have no choice other than a public shelter. That's your last resort. Don't expect more than a cot and a restroom. Make plans for your pets. In the worst case of shelters filling up, you may not even have that option.

The cliché "Run from the water, hide from the wind," holds true in many locations. It means if you are in a flood zone or storm surge zone, you have to get out of the way of water. If you are not in a flood zone, but in a solid building, you can often shelter in place.

The impact of tropical storms and hurricanes is greater now than it's ever been because more people live along the coast and more people own more property and more expensive property than ever before. This means a weaker hurricane striking a highly populated area can have much more negative effect than a strong hurricane would have had decades ago striking the same area with low population.

In long-term planning, homes in hurricane zones must be designed for the risks of flooding near water, by elevating them in flood and surge zones. Existing homes and businesses can benefit by installing flood protection systems. These may be walls or panels. Whether it's rising water from storm surge or rising water from extreme rain runoff, flood barriers mitigate or prevent water intrusion.

High wind threat requires solid construction. The roof style and windows have to be engineered to minimize wind and rainwater impacts. Basic window coverings like shutters make a lot of sense. That's why you find them on many older homes in coastal communities. Those are shutters that actually can be closed, unlike some newer homes that have ornamental, non-functioning shutters. Shutters only protect windows. Every component of a building's structure from the roofing, to the nails and fasteners, to the support walls and foundation, must integrate into a fortified unit. Some new homes that are not in surge and flood zones are being built of concrete, reinforced with steel bars. Not only are these homes able to stand up to very high wind, but they are also more fire-resistant, termite resistant, and more energy efficient. Concrete homes may cost a bit more than most homes initially, but they cost less to heat and cool annually, and homeowners may also save on insurance premiums.

If you've been impacted by a hurricane, you know these things: Spoiled food. No air conditioning. Highway bridge destruction and the inconvenience of long alternate routes for months. Businesses close to never reopen. Insurance headaches. Contractor hassles. Uprooted families. Long-lasting emotional scars. On the other hand, most storms bring out the best in people, helping people. When you live on the edge of the tropical arena, natural occurrences like those become human disasters, when caught in the midst of nature balancing air, moisture, and heat via whirling monsters. They are human disasters because we do have some control over the impact. Personal and community planning and building to mitigate the perils of a storm is an investment that pays dividends in life, lower stress, and in dollar savings.

The next big hurricane will surely produce similar impact, with survivors experiencing loss of satellite TV, internet, and cell phone service. What will we do without social media?! Our interconnected digital world will remind us how some of us have lost basic survival skills, like knowing your neighbors who become first responders when you are in trouble, and vice versa. The next big one will also be more costly than similar storms in the past simply because population and development on the coast are so much higher than ever before, and because we are a fairly rich society. Look at the size and cost of beach homes compared to what they were 50 years ago! We have more stuff, like multiple vehicles and multiple TVs in our homes. We have so much stuff that we pay to store the extra stuff! Many of us carry a phone that costs over $500.

Could your grandparents have conceived of that when they were young?!

Right now, on your expensive phone, document all your possessions with photos, and a written inventory list. You could lose everything in a windstorm, flood, or fire. Make sure your insurance covers the threats specific to your neighborhood.

Even for a medium hurricane or a strong tropical storm, inconvenience and disruption will grow because we all try to cram so much into a day and week, based on a digital clock, not an analog atmosphere. When the next big one comes, you'll be reminded of the value of your life, your health and your comfort. That goes for family, friends, and pets too. Have a plan!

Hurricane hunters

The United States is unique in forecasting hurricanes because of fleets of aircraft dedicated to going into these storms for data. One is the U.S. Air Force Reserve Hurricane Hunters, based at Keesler Air Force Base, in Biloxi, Mississippi. Their routine mission is to fly a C-130 aircraft into the storm, both day and night when needed. Flights become more numerous and frequent when storms are likely to threaten cities, states, and countries.

I've had the pleasure and adventure of flying with the Hurricane Hunters. Media are able to fly with U.S. Air Force Reserve Hurricane Hunters by request, with limited space availability. While they are very active, often with day and night daily flights when a storm threatens the U.S., I become very active, too, as a broadcast meteorologist. If a storm is pointed toward my area, I can't and wouldn't go. Likewise, if a storm threatens Keesler A.F.B. I wouldn't go because they have to move planes to a different base. Even if they were to take off from Keesler for a mission, they might have to land in another city. The second regular base of operations is in St. Croix, which handles storms in the central Atlantic.

The hurricane hunters fly a standard C-130 propeller plane, outfitted with only a little extra weather sensing gear. It is a durable and powerful aircraft that does well in the heavy rain of a tropical system. A typical flight crew is a pilot, co-pilot, and navigator, and the technical crew is a loadmaster who drops weather instruments into the storm, along with a meteorologist who takes visual readings.

Basic weather gear is an on-board radar, a device that measures sea surface conditions using microwave signals, and the dropsondes—the instruments dropped into the storm at multiple points to measure temperature, humidity, pressure, and wind as they fall slowly with parachute, to the ocean. The dropsondes weigh a couple of pounds so they are only dropped over water, where they send data until splashdown. They are not recovered. Data is sent back to the National Hurricane Center in real-time.

Even in the strongest hurricanes, the steady wind is not an issue for flight. When you consider that in still air, a plane may have a ground speed of several hundred mph, a headwind or tailwind of even 200 mph only changes the ground speed. A head wind would lead to faster fuel usage. In most cases, the plane has a crosswind, and that requires "crabbing" where the plane points somewhat into the wind but is blown sideways, still able to maintain a straight-line path.

The National Hurricane Center tasks the Hurricane Hunters with gathering data. This is done for any storm that is a potential threat to the U.S. and neighboring countries. For media guests on the planes, the flights require a release form that states the potential risk of flying into a storm, and acknowledgement that the plane may be diverted mid-mission for some other military purpose. That, and the fact that the planes do not have lifeboats, is sobering.

I've only done daytime flights, because I like to look out the window. Those flights typically take off around sunrise, requiring a pre-dawn arrival to Keesler A.F.B. Entering the base, you gain an immediate respect for the armed service members, simply because they are armed. You also quickly learn to follow instructions and not cross lines, literally. Lines on the tarmac are to protect you from danger of things like propellers, and to protect the military assets from humans. The flight crew holds a briefing for each other to determine the approach and flight pattern. In weak systems, with little turbulence, data collection may occur at a 5,000-foot flight level, or even lower. In stronger or turbulent storms, the flight level is closer to 10,000 feet, for safety.

My first flight was in 2007, into Tropical Storm Gabrielle. The storm was so close to land that our flight was generally within just half the storm. My second flight was the next year, for Hurricane Ike, in the central Gulf of Mexico. It was a Category 2 hurricane that would later strike Houston, Texas. My third flight was into Hurricane Nate, in the southern Gulf of Mexico, in 2011. It briefly was a low-end hurricane. What made that my most memorable flight is that the two flight meteorologists on board were both former students of mine at the University of South Alabama.

The mission flight is typically 8 to 10 hours. In the case of Hurricane Nate, it was a couple hours getting to the storm from Keesler; around 4 hours inside the storm; and another couple of hours to get back to the base. In the storm, the C-130 plane flies a repeating pattern of crisscrossing the storm from quadrant to quadrant, passing through the center each time, about 2 miles above the water. The majority of the time in the storm you are in clouds and can't see much, but in the eye, you can see the ocean and surf. The flight meteorologists take instrument readings and also visual readings. With the rank of captain, my former students showed me how they do their jobs, estimating wave heights and collecting data, and asked me for my opinion as a fellow meteorologist!

Unlike what most people might imagine, the majority of the flight is not turbulent. While the plane tries to fly in a straight line, it does deviate slightly, for safety. The navigator uses on-board radar to help the pilots avoid the most turbulent thunderstorms. On this flight, the only time I wore a seatbelt was during takeoff and landing, and during the few minutes when we would cross into and out of the eye. Even then, the turbulence was no worse than what you feel in a commercial airplane. Most people would probably be bored on the flight, but it is fascinating to a meteorologist.

While the plane is large, there are only a few windows on each side, so the view is limited. From time to time, pilots may allow media in the cockpit for a forward view. In the long stretches of travel in the clouds, or outside the storm, one can stretch out on the bench-style seats that line the left and right side of the plane. They are not comfortable, but after a night of abbreviated sleep, most people can handle it. There is a basic toilet behind a pull curtain in the back of the plane. Some planes may have a small microwave, too, but I wouldn't try to eat a full meal. While turbulence was minimal on all the flights I took, if it gets bad, your stomach wouldn't be happy. My meal before the flights was light, and consisted of things I know wouldn't upset my stomach.

What surprised me about the flights? How much power the plane has when taking off. How smooth most of the flights were. Seeing ships traveling in the eye of a hurricane. Seeing how even the bands of a hurricane have layers at different altitudes. It's not just one solid cloud object.

For the crews, most of whom are Reserve Officers, the flights are routine with little more risk than other aviation. They provide unique and detailed data that cannot right now be effectively gathered any other way.

The National Oceanic and Atmospheric Administration also has hurricane hunter aircraft that fly into tropical systems. Their mission is more research for long-term improvement in forecasting even though data is used in real-time. NOAA P3-Orion airplanes fly at a higher altitude to sample more of the storm and nearby environment, and a NOAA Gulfstream jet flies even higher to gather data.

There's a newer high-flying aircraft that investigates tropical weather. It's NASA's Global Hawk. It is an unmanned plane or UAV or drone. NASA outfits these research aircraft with radar and hundreds of pounds of sensing equipment and dropsondes. Not only do the NASA Global Hawks measure weather high above the sea surface, but they detect and measure particles like dust from Africa to see how that impacts tropical systems. The strength of the Global Hawks is they can stay in the air longer than other aircraft, to reach greater distances and gather more data.

Tropical weather forecasting

When tropical weather systems get within a few hundred miles of land, radar is finally able to detect the rain within the storms. Radar data is definitely used in forecasting. Computer models project where a tropical weather system will be in time, and how strong it may be. Data from satellites, radar, buoys, weather balloons, and hurricane hunter aircraft are fed into forecast models. These predictions have substantially improved over the years for location but not quite as much for intensity.

The National Hurricane Center uses computer models, historical data, satellite data, hurricane hunter data and current weather maps to create a forecast cone. Initially, forecast positions are determined with a radius of average error. That gives a series of circles, for each time period, larger with time, since error grows with time in any forecast. It is the same for a pitcher trying to hit a target. When the target is close, the circle of error is small. As the pitcher backs away from the target, the error circle grows larger. Rarely can a pitcher hit the exact center each time. For tropical weather forecasting, the edges of the error circles are then connected with a line to create a cone. The cone shows where the center of the system is predicted to be. Because it is a projection, there's no guarantee that the center will stay within the cone. Just like a pitcher cannot always throw a strike and stay in the strike box, for tropical storm forecasting, 33% of the time the center may go outside the circle or cone.

What's maybe more important to know is that even when the center of a storm stays well within the cone, the impact of the storm, in terms of surge, flooding rain, and tornadoes, can easily extend beyond the cone, especially on the right side of the path. A hurricane is not a single point on a map. The forecast is only as good as the current understanding of tropical weather and the ability to accurately measure and simulate it.

We know that hurricanes are driven by warm water evaporating by heat. When the water condenses, it releases latent heat, powering the storm. Low pressure draws air inward in a counterclockwise spiraling pattern that forms bands of rain. The air and rainbands converge near the center, increasing speed. Dry air sinks in the middle of the storm, creating an eye. Around the eye, air rises in cumulonimbus towers. At high levels, rising air and cirrus clouds are evacuated from the hurricane by high pressure blowing clockwise.

The central pressure of a storm is a good way to estimate the overall strength. Usually, the lower the pressure is, the stronger the storm is. When pressure goes down, wind has to increase to keep the air mass in balance. Think of the low pressure as a valley, and the wind is a boulder rolling downhill. The lower the valley, the faster the boulder moves. We humans can't easily sense atmospheric pressure but some of us do feel rapid pressure changes in our joints. Everyone feels wind, and wind creates storm surge and property damage, so that's why meteorologists tend to put more focus on wind than pressure. A 100-mph wind is the same in any storm but a pressure of 28.10" can mean different wind in different-sized storms.

When the impact of a hurricane or tropical storm is possible within 48 hours, a watch is issued to give people time to plan and prepare. When the threat is likely within 36 hours, a warning is issued for the areas most likely to take the direct impact. These watches and warnings are specific to what the storm will likely be and what the threat may be. They don't always match the current status of a storm, and that confuses people. It's most confusing when there's not even a designated storm. This is why you now hear of potential tropical cyclones. That designation is used when there's no named or numbered storm, but one is highly likely to spin up quickly and make a rapid landfall. In order for governments to issue watches and warnings and for locales to activate tropical storm and hurricane plans, there has to be something designated, even when nothing exists! Pre-designating a future storm as "potential" is the best solution to raising awareness to a threat that will seemingly come out of nowhere.

Subtropical is another term that's new to the public. It's when a tropical storm is a hybrid between a tropical storm and a regular low-pressure system. It's a minor technical distinction that does not change the impact of the storm if it hits where you live. A tropical storm has no fronts, while the highest winds are close to the center. A subtropical storm may have fronts, with the highest winds usually more spread out from the center. Both start over water, and either can grow to become a hurricane. On satellite, to the untrained eye, they are similar. It's like comparing an electric engine car to a gasoline engine car. On the outside, they look pretty much the same, but under the hood, they are very different. In a collision, the impact is identical to the passenger.

The worst part of many hurricanes and tropical storms is not always the initial impact. It may be the flooding and tornadoes that linger for a few days. It can also be the physical, mental, and financial stress of clean up and recovery. That may be days and weeks without an intact home, electricity, clean water, ice, air conditioning, internet, fresh food, or income. For all of the disruption and destruction that tropical weather causes, it is a vital ingredient of weather that transports energy and moisture between sea, air, and land, around regions of the planet.

Hurricane records go back to the middle 1800s. Prior to the launch of geostationary satellites in the 1970s, undoubtedly there were tropical storms and hurricanes that did not make landfall and went unnoticed. For that reason, hurricane records are fuzzier and dustier, the farther back in time you look. Old records don't have as much fidelity as recent records.

If only a hurricane record were just a vinyl disk, used by a D.J. to entertain and bring joy to the masses. No, a hurricane record is an extreme of some sort, where the most memorable records are often not pleasant. In recent years, we've seen records for early storms, intense storms, landfalling storms, and for total numbers of storms in seasons. That's why we all look forward to the last calendar day of hurricane season, like an 18th birthday. I hate to have to remind everyone that the calendar does not control weather. Similarly, local weather does not control the formation of tropical tempests, although it does play into what a storm can do when it approaches a specific area.

Hurricane trivia

Tropical storms have happened in EVERY month of the year, in the Atlantic Ocean. What about hurricanes? The only months in which a hurricane has never been recorded in the Atlantic are February and April, according to the Atlantic Oceanographic and Meteorological Laboratory of NOAA (National Oceanic and Atmospheric Administration). In the colder months, tropical activity is rare. Fortunately, outside of the calendar hurricane season, for the United States, there has been no documented landfall of a hurricane.

Hurricanes, at a minimum, are an inconvenience. They throw us off balance. Way off balance. If there were a way to stop a hurricane, would you? Who would get to make that decision? Who would pay for it? Would we stop a hurricane that goes to only certain areas? How about stopping one on Labor Day and scheduling it for another day, like Tax Day? What about the beneficial rain that regions in drought need?

I've heard many ideas on how to stop a hurricane. Nuclear bombs are out of the question for the radiation fallout, but also because one single afternoon thunderstorm has more power than a nuclear bomb. It's just released over dozens of minutes rather than over a few seconds. Given that hurricanes are made of thousands of thunderstorms, you should quickly dash that thought.

People suggest towing icebergs to warmer water, adding chemicals to the ocean to limit evaporation, or adding ice or chemicals to the clouds to slow their growth. For a detailed description for why these are scientifically not reasonable or feasible, do a web search for "NOAA AOML Hurricane FAQ."

Of the many theories proposed to stop hurricanes, each comes up short and each would have unintended consequences to the environment and to life. Each would have gigantic cost, and each would require an enormous amount of energy. Each would be a gamble. Each would have to be decided upon by a human and then litigated by another human when harm befalls more humans.

We feel the inconvenience and destruction of hurricanes, but hurricanes are just as vital to the planet as volcanoes are. Would you try to stop those from erupting? Maybe pouring cement inside of them would help. Nope, that's like trying to totally stop yourself from sneezing. You can't. We should not try to stop hurricanes; they are a natural part of Earth maintaining balance. A hurricane redistributes heat, moisture, and electricity in the air. In the ocean, it moves heat, too, and salinity. Hurricanes also rebalance concentrations and locations of dust, seeds, nutrients, trees, and some birds and creatures. Plan and prepare for the perils where you live. In the classic words of a 1970s commercial for Chiffon margarine, "It's not nice to fool (with) mother nature."

There is always high interest going into hurricane seasons for long-range outlooks. One of the earliest ones that is widely publicized comes from Colorado State University. Some people question how a university at the foot of the Rockies can make tropical weather projections. It's not hard. We forecasters are trained for all weather types, and worldwide weather data is available by computer. With long-range outlooks, like those for hurricane season, some people chuckle about an outlook or forecast update being easier in the middle of a season than it is at the beginning, but that's no different from a GPS update in the middle of a trip, or a coaching change in the middle of a game. In everything in life, a wise person adjusts their outlook when new data shows changes to a likely path.

More than a half-dozen agencies, universities, and businesses in the U.S. create hurricane season outlooks and release them to the public. Aside from Colorado State, you'll find outlooks from The Weather Company, North Carolina State University, AccuWeather and NOAA. Many more are done outside of the United States. Outlooks have been done for decades but now, with the internet and social media, you can't escape them. There are many more projections from other businesses that are not made public. Are any of them any good? Yes. All seasonal outlooks have value but not for your house! Outlooks give a statistical likelihood of over, under, or average, but they are not a specific forecast.

Hurricane season projections generally cover the entire Gulf of Mexico, Caribbean, and Atlantic Ocean—a total area that covers somewhere between 15 and 20 million square miles. This makes it hard to compare different outlooks. Most are not for the odds of landfall or how long tropical storms or hurricanes last. Most don't specify where or when named storms are likely. They are just for the total number. Who gets value from these types of outlooks? Industries, agencies, organizations, and businesses that have many assets in coastal locations, because the outlooks are for regions, not specific cities. With multiple locations, statistics are helpful. Think big box stores, utility companies, convenience stores, and our military.

Practical hurricane threats are only determined week by week, based on air pressure, steering winds and water temperature. Understand that an above-average season includes seasons when there are a bunch of storms that stay in the Atlantic and don't make landfall. It could be many brief, low-end tropical storms and hurricanes. A below-average season can have just one major hurricane that makes landfall where you live, as a historic event. An average hurricane season can set up a pattern where several hurricanes strike the same general area in a matter of months. None of those is a prediction. Each is a scenario to help you understand why seasonal forecasts are not gospel.

Prepare for every hurricane season the same way—have a plan and the tools to get you through the storm. Don't follow myths. There's no specific pattern that follows an early-season storm. It's not unusual to get a named storm in May. Don't let random people's opinions and perceptions of weather worry you. Just follow the forecasts from a source you trust, like NOAA. Know the terms and vocabulary used to describe storms and threats.

Fujiwhara, "What the Fujiwhara is that?!" The name, "Fujiwhara," entered social media musings of tropical storms Marco and Laura merging into a bigger storm in the Gulf of Mexico, in 2020. It didn't happen, and it rarely does, but a century ago, Dr. Sakuhei Fujiwhara described how two large spinning objects could pivot around each other. That's known as the Fujiwhara effect.

Hurricane stages

It's sometimes easier to predict human nature than nature nature. Here's my human nature forecast for the 10 stages of hurricanes:

Stage one. Denial. "That's not going to develop. I'll ignore it. The last one didn't do anything so this one shouldn't do anything either." That's short-sighted.

Stage two. Social media speculation. It's like football season. Everyone gives an opinion on what's going to happen. "Should I be worried?" Don't be scared, be prepared.

Stage three. Forecast cone confusion. "Are we in, are we out, why does it keep moving? I don't understand the cone." The cone is only for the forecast center of the storm, not the impact.

Stage four. We're under a watch. People become concerned. Storm veterans go into preparation mode. "What should I do? Should I evacuate?" It depends on your proximity to beaches and surge and flood zones and what type of structure you live in.

Stage five. We're under a warning. Rush to store and gas station. Some people surf TV and social media for a forecast that makes them feel better. Some are oblivious to the threat. Others compare forecasts and ask me which one is right. Mine, of course. Why else would I give it if I didn't think it was right?!

Stage six. Impact. Disruption. Hold on.

Stage seven. Surprise, disgust, and dismay. A few say it was no big deal. Many say, "No one told me it was going to be that bad. I'll never ride one out again." At the same time, forecasters are criticized for not saying how "bad" things would be, and criticized for making the threat seem more than it was. Different impacts for different locations.

Stage eight. Survival and repair. The phrase, "The first 72 are on you," is recognized for what it means. For the first 72 hours after a hurricane, you can't expect someone else to help you. You have to be self-sufficient, and then hope for speedy insurance claims and maybe assistance from the government.

Stage nine. Repair and recovery. Face the challenge of finding reliable contractors, building materials, and getting utilities and infrastructure back in place.

Stage ten. Wait and prepare for the next storm.

There's never a blanket answer to, "What should I be doing?" when you are in the forecast cone. Your planning will be determined by how vulnerable you are. If you are on the coast, in a flood zone, in a mobile home, with no property or flood insurance, you are at far higher risk than someone with full insurance who lives in a fortified home, with a generator, on high ground, with no overhanging trees.

When you are in the fifth day portion of the cone, pre-plan the "what if" and have a checklist of all the things you might need to do to handle the worst-case projection. Casually, do the daily little things that you might do anyway, like getting gas, food, prescriptions, and cash. Avoid the rush and panic by people who wait too late.

In or near the cone, four days away, definitely have a plan. Start thinking about rescheduling appointments and travel, even if they are for shortly after the day of possible impact. If there are things you need to purchase, do it now, to beat the crowd.

Three days away, in or near the cone, watch the forecast more closely for details of potential threats. Have a plan for if you have to evacuate. The more things or people you might need to evacuate with, the more quickly you may have to go into action, especially if your community has dense population to clog roadways.

Two days out is when watches are issued to outline wind and surge threats. Stop focusing on the cone. Focus on watches. If you are in a watch, take action, whether that is boarding windows or preparing for rising water. Warnings are issued a day and a half before impact to narrow the threat area. Wash clothes and dishes in case you later lose water and power.

One day out, you should have completed all preparations. Plan for electricity loss by keeping devices charged. Listen to details of possible impact of wind, rain, storm surge, and tornadoes. Hope for the best but plan for the worst.

Never before has technology given us such precision. When I was a kid and someone asked, "What time is it?" you never heard 9:13 or 9:17. You heard, "quarter after nine." If my parents told me to be home at 7:30, that meant I could be home by 7:40, maybe even 7:45 and not be in trouble. Now, with computers and pixels we look at everything in the smallest quantity. That's called precision. Precision is simply the amount of detail. If a chef puts 1.26 cups of water in a mix, then that is a very precise amount; but if the recipe called for 1 cup of water, then the chef was not accurate.

With tropical weather, you see precision in the forecasts down to the pixel on your HDTV or widescreen computer or on your phone retina display or on your color laser printout. This can mislead you to believing that the future can be forecast with super accuracy. In fact, it's not easy to measure tropical storms with high accuracy. Much of what we gather about tropical storms is a bit fuzzy.

Find your favorite old family photo from a couple of generations ago. It's probably a bit blurred or fuzzy. Scan it and blow it up on your screen to 1900 pixels wide. You now have more pixels showing you the same picture, but the image is no clearer. It's just bigger on a prettier display. It can't be any clearer than the original image. There's no new detail in the picture. If you step back now, and look at the big picture, it becomes clearer, but never perfectly clear. This is where we are with distant tropical systems — they are too far out for an accurate landfall forecast. Precision and accuracy are not the same. The portion of the five-day forecast cone that first touches our coast may be over 400 miles across. The closer any storm gets, the smaller the cone gets, even though the impact can and will also extend outside of the cone.

Keep a healthy and realistic perspective of all weather forecasts, where uncertainty is ever-present, and the future is uncertain. Walk into a maternity ward, pick out the cutest baby and ask the nurse, "What do you think he will become later in life?" Ask the parents, "Where do you think she will live when she grows up?" You may either get a chuckle or a blank stare. Predict the birth of a baby two days away. Even if you've had one or two children already, that tells you nothing about how and when the third will arrive and what its personality will be. You're only guessing and assuming hair and eye color, weight, and arrival time. Good luck with that. We can't accurately predict our own futures so it should not be a surprise that we cannot accurately predict nature.

When a storm seems to be targeting you, "Should I ... move my boat, board my windows, evacuate"? Given what we don't know about the future, take what you are hearing in the forecast, speak to others in the same situation or community as you, get their best advice for what has happened in the past and then do what lets you sleep well at night. If you have the resources to take extra precautions in any hazard, you're generally in a better position. There is rarely a yes or no answer to "should I" since each of us is in a different situation and the weather in every neighborhood will not be the same.

What you shouldn't do is worry or panic. You shouldn't put too much faith in chatter online from unknowns who present themselves as experts. You shouldn't spend too much time listening to people with visions of doom. You shouldn't think that every change in the forecast or coordinates or status of a tropical system means the outcome will be different. No matter what anyone says, thinks, or guesses, weather occurs without any consideration to human thought or population.

Plan wisely, prepare. Have a backup plan. If a storm misses you, don't be surprised to see a ripple effect that does reach you, like loss of electricity or cell service. Infrastructure is strong in most communities, but it only takes one failure point to inconvenience thousands or millions of people.

New and distant tropical systems have many possibilities. We watch day by day, and so do millions of others. Each of us looks at a hurricane like we are the basketball rim, and the hurricane is the ball. Five days away the shooter is under the other rim. Four days away the shot comes from past the center court near the other free-throw line. Three days out the shot is from half-court. The day before it is from the three-point line and the day of landfall it's right from the free-throw line. "Miss, miss," is what we say, and so do the millions of others up and down the coast.

You are not the only possible target for a hurricane, but the final shot is a layup. That's when time is up. Are you ready for whatever may come? Take it with the training, preparation, and grace of a champ, knowing there will be winners and losers, and more hurricane games to play, although they are never fun games. In a hurricane game, tropical storms and hurricanes can strengthen or weaken, speed up or slow down, veer left or right, stall, loop, or all of the above. Most travel at a constant speed in a smooth curve or line but that doesn't mean they all will. Some make landfall as hurricanes that are rainy or not so wet; windy or not so windy; with large surge or little surge; and with many tornadoes or no tornadoes. From these four areas there are dozens of combinations of how a hurricane can make landfall. Given potential rapid cycles of growth and decay and how little detailed data we can gather over the oceans, we end up with large uncertainty, especially days in advance.

Hurricanes are not a single point. The landfall impact covers a region. Within that region, depending on what type of storm it is at landfall, the impact for those along a river or bay will differ from those on high ground. We think of the coast as taking the full force of a storm, but inland communities can suffer flooding and tornadoes. Do your basic hurricane planning before the threat, but always be ready to deal with a big change. Don't rely on one single parameter, like the category of a storm.

Predicting the impact of a hurricane landfall is like predicting the outcome of a two-car collision. The speed and the size and weight of each vehicle are big factors, but so is the angle of collision, the age and design of the cars, and whether they have air bags or not. Critical to the outcome is whether the passengers are wearing seat belts.

A hurricane impact forecast leads to the issue that a storm that is meteorologically identical to another will have widely different impact in different regions, and in different times of the year.

Hurricane ACE

When you hear hurricane ACE, you might envision the name of a first storm of the season. It's not. You might think it's the nickname of a hot-shot hurricane hunter in a Tom Cruise movie. ACE is actually an acronym used by meteorologists. It stands for Accumulated Cyclone Energy. ACE is the total energy expended by a tropical storm or hurricane, another way to quantify storm strength and activity in the tropics.

The 2021 hurricane season produced 21 named storms, of which 7 became hurricanes and 4 of those became major hurricanes. It wasn't a record, but that above-average count made it very active, just like the previous 5 hurricane seasons. Four of the 2021 named storms lived for less than 48 hours and did not strike land. Decades ago, they may have gone undetected. Brief and often weak storms can skew the total season count upward.

Bring on ACE. Accumulated Cyclone Energy goes beyond the simple counts of how many tropical storms and hurricanes occur. ACE accounts for the longevity and intensity of all tropical storms and hurricanes. It's as simple as taking the wind speed of a named storm at every 6 hours, squaring it, and then adding all of those values together. ACE gives a better accounting of the total power of a storm, when, for example, the system lives as a minimal tropical storm for a week, briefly reaches hurricane strength, and then fades. It also accounts for a short-lived hurricane that reaches category 5 status for just a few hours before quickly dissipating. Shorter-lasting and minimal strength named storms don't contribute much to ACE, making ACE a good way to compare recent storms and seasons to those in the decades before satellites and computers.

Using a measure of Accumulated Cyclone Energy, the 6 hurricane seasons through 2021 would still be more active or more energetic than average. However, when you look at the record 30 named storms in 2020, that season turns out to be only the 13th most active hurricane season, in terms of ACE, according to researchers Eric Blake and Chris Landsea. Yes, Landsea really is his last name.

While total storm numbers have increased in recent decades, ACE has not increased as much. ACE is not a perfect index, but it is one of several. Total energy relates more to potential impact and damage than does just a peak wind. ACE does not tell you about landfalls, and it certainly doesn't give clues as to whether a tropical depression can sit and flood a region. A tropical depression that never reaches named status will not contribute to ACE, but it can still be costly and deadly.

What ACE shares with the traditional method of simply counting total numbers of named storms is that seasonal predictions of either do not give you the answer of, "What will happen here?!"

Hurricane retirement

Where does a hurricane go for retirement? Into history books. World Meteorological Organization policy is to retire storm names for tropical storms or hurricanes that are devastating in number of deaths and/or in dollar damage. Keep in mind that more and more hurricanes that make landfall in the U.S. are multi-billion-dollar storms that take lives. Those hurricanes are more impactful because there are so many more of us along the coasts, with so much more expensive stuff. Over 90 tropical storm and hurricane names have been retired since the 1950s. The WMO criteria looks at total impact over multiple states and countries.

When a hurricane name is retired, a backup name is inserted into the rotating annual list of names. In hurricane seasons that go beyond the standard list of names, there is now an overflow or standby list of names available.

CHAPTER 9

Sky Sights and Optics

What's a bow that can't be tied?
A rainbow.

M eteorology has a foundation in observation. We can't predict the weather until we observe and measure it. Beyond the scientific value of observation, the reward is seeing and experiencing beautiful and intriguing and sometimes rare phenomena generated by the interplay of light, moisture, and air.

Light travels through transparent objects like ice and water. In the process, it slows down slightly. That causes refraction, the bending of light, which can separate it or disperse it into the seven colors of the spectrum. When light passes over tiny objects, its path can shift slightly. That is called diffraction. Commonly, when the path of light is through a cloud of particles like dust, pollen, pollution, and other aerosols or even water or ice of a natural cloud, it is softened in the process of diffusion. Diffusion, refraction, diffraction, and reflection make things interesting. Cloud color and sky color give science clues as to how light travels through air, and about what type of particles are floating in the air.

The scattering of light by what's in the air makes rays of light visible. Scattering also puts color in the sky. Blue is the color of the sky, a clean sky. From the spectrum of color arriving from the sun, the shorter wavelength colors are scattered more by air molecules than the longer wavelength colors. Blue photons are scattered the most. Once you factor in moisture and aerosols and how high the sun is, we enjoy a sky that can range from pale blue to royal blue. Looking lower along the horizon, the blue becomes more pale or white because the thicker layer of atmosphere scatters more of the longer wavelength colors to dilute the blue. Longer wavelength colors like red, orange, and yellow are increasingly scattered in the troposphere when the sun is low and light passes through more air. Without the air of our atmosphere, the sky would be black, as it is in space.

Sky color and cloud color are also tinted by city lights at night. Manmade light adds a glow over large cities, known as light pollution. Low clouds reflect light back to the ground so that on cloudy nights, cities may be brighter than on a clear night. If streets are wet or when there is a fresh coat of snow on the ground, even more light will be reflected upward to brighten the night clouds.

The electric discharge of thunderstorms can show color too. Lightning is generally white, but the color is shifted by moisture and aerosols in the air between you and the lightning stroke.

Large aerosols, like sand and dust in dust storms, easily change the color of the sky to the color of the particles. When those aerosols are dense enough, they dramatically dim sunlight, and they may lower visibility to dangerous levels for vehicles and aircraft.

The interaction of light with clouds creates pretty and fascinating scenes. Any thick cloud that reflects sunlight back to you appears white and bright, but when viewed from the other side, that same cloud seems dark and ominous. When you are in the shadow of a cloud, it means the cloud is thick enough to block the sunlight from getting through. Around the edges of the cloud, the light does pass through, creating a bright highlight or lining. That probably explains the origin of the phrase, "every dark cloud has a silver lining."

When light passes around gaps in cumulus clouds that fill the sky, we see rays of sunlight that seem to fan out. These are known as crepuscular rays. The beams of light are actually parallel, but they always appear to start at a single point and then spread out. This is no different from how rows of crops seem to meet at a distant point or how train tracks seem to converge in the distance. Crepuscular rays produce the opposite effect of cloud shadows. Instead of only seeing a darker area produced by a cloud, we see lighter areas where breaks in clouds allow light to shine through.

Funnel clouds similarly show color based on the position of the sun compared to adjacent clouds. The moisture condensed in a funnel cloud in direct sunlight will be bright and white. Funnels in the shadows of other clouds will be darker. Once a funnel cloud touches water or land, its color will come from whatever it picks up in the circulation.

Clouds reflect the colors the sun sends, so when the sun is low in the sky, clouds become yellowish orange, to sometimes red or pink. When the sun is below the horizon, it can still shine upward to high clouds, illuminating their bases, highlighting their texture, and allowing them to send shadows upward, through the air or onto even higher clouds.

"Red sky at night, sailors delight," is an old expression, with science behind it. It refers to often breathtaking sundown displays of longer wavelength colors, like red and orange, in the air. When the sun is low in the sky, the light passes through a greater amount of atmosphere, and this scatters out the shorter wavelength colors. In order to see a red sky near sunset, the sky to the west has to be mostly clear. In middle latitudes, weather tends to move from west to east, signaling that the incoming weather will be clear. That would make a sailor delighted!

Rainbows usually produce happy feelings too. I have never seen anybody upset to see a rainbow. Maybe it's in our DNA to be overcome by feelings of joy when colors arc across the sky. There is some science behind rainbows and happiness. In middle latitudes, rainbows in the afternoon generally appear when the rain moves away from you to the east, and the sun comes out in the west. An afternoon rainbow signals improving weather, when the weather pattern is moving from west to east. When the rain tapers and the sun returns, low in the sky, somebody can see a rainbow if they are in the right spot. Look eastward in afternoon in the direction of your shadow. The angle of the rainbow depends upon the angle of the sun. The lower the sun is, the higher and wider the rainbow is. Natural rainbows can be seen when the sun is lower than the angle of 42 degrees. Morning rainbows are seen in the west up to several hours after sunrise.

The brightness of a single or double rainbow depends upon the size of the raindrops and the intensity of the sun. Even though a YouTube video years ago made double rainbows famous, they are common and captivating. Just look. Rainbows form when sunlight shines on a curtain of raindrops. Light enters each raindrop, refracts on the way in, bounces on the inside of the raindrop and then refracts on the way out. The refraction, or bending of light, separates it into its component colors of red, orange, yellow, green, blue, indigo, and violet. The result is an arc of color seen in the direction of your shadow. The higher the sun is in the sky, the lower the rainbow is and the less of the rainbow you see. The center point of the arc of a rainbow is actually centered on the shadow of your head. In the middle of a summer day, the only rainbow you can see is a manmade one, unless you are far north of the equator where the sun is lower. From an airplane, you can sometimes spot a midday rainbow, too, beneath the plane. Similarly, you can see a midday rainbow if you are high on a mountain or skyscraper, looking down, toward your shadow. It can actually be a full circle.

As the sun gets lower in the sky, and we are at ground level, the rainbow gets higher and wider. When you look closely at a rainbow, you'll notice the color red on the outside, with violet on the inside. This is the primary rainbow. It's not unusual to see a fainter second rainbow outside the first one. In the secondary rainbow, the order of colors is reversed. It's less bright because the light takes an extra bounce inside the raindrops. You may also notice between the primary and secondary rainbows is a slightly darker area. While the sky is a little brighter just outside the secondary rainbow and also brighter inside the primary rainbow, there is sometimes a repeating pattern of purple and blue just inside the primary rainbow. These repeating colors are called supernumeraries. All of this is simply how light interacts with water droplets.

The brighter the sun and the larger the raindrops or the more raindrops there are, the more brilliant the rainbow appears. Rainbows are useful in weather prediction. A rainbow seen in the morning means there is rain to the west. In middle latitudes where weather tends to move from west to east, a morning rainbow suggests wet weather moving in. Afternoon rainbows are seen in the east but only when the sun is shining in the west. Afternoon rainbows appear when the rain or storm is ending or moving away. Perhaps that's why rainbows evoke positive emotions. There are exceptions to almost every rule. A rainbow can appear during the middle of a storm while there's still a threat of lightning.

Refraction in ice crystals creates another spectacle that is less noticeable but not uncommon. It's called a halo. Halos are less noticeable than rainbows because they happen above the horizon and overhead, rather than directly in front of us. A halo occurs around the sun or moon when there is a thin layer of cirrus clouds. Light passing through the ice crystals of cirrus clouds refracts or bends and is separated into the seven colors of the spectrum. Unlike the rainbow, the color red is on the inside of the ring of the halo. The halo happens in the direction of the sun, while the rainbow is found in the direction of your shadow.

There are also many accessories to halos, some of which are fleeting, while others linger longer. Where and how they happen depends on the type of ice crystals — whether they are plates or columns or needles. It also depends on whether most of the ice crystals are aligned the same way or tumbling randomly. Bright spots on either side of the sun are common with halos. These are called parhelia which means, "with the sun." The nickname for parhelia is sundogs or mock suns. Parhelia could be broad and colorful, or concentrated with not much color.

If the halo is high in the sky, then there may be an additional larger but fainter halo around it. Other times, we find a bright portion of the wider halo beneath the main halo. This is called a circumhorizon or circumhorizontal arc. A circumhorizon arc shows up as a medium, and sometimes long, straight segment that you can only see if there are enough cirrus clouds beneath the sun. They don't happen as often as halos do. From refraction, or the bending of light in ice crystals, the seven colors of the spectrum appear. The colors always line up in the same order, with red on top.

When you stop to look at a halo around the sun, known as a solar halo, you quickly find bright sundogs on the sides, along with an additional arc on the top of the halo, known as a tangent arc. Between the sundogs is yet another bright arc that forms a complete circle in the sky.

This larger circle seen in the cirrus clouds is known as a parhelic circle, since it passes through the sun. Par means with and helios is the sun, so a parhelic circle is something you find with the sun. Ice crystals create all of these spectacles.

People who look straight up when a solar halo is in front of them may be rewarded with the sight of another inverted arc of color that is centered directly over their head. That is called a circumzenithal or circumzenith arc. It's a colorful arc around the zenith in very thin cirrus clouds. The circumzenithal arc is not a rainbow, but it is a rainbow impersonator, with the same 7 colors. It's way high in the sky, directly above your head, and it forms when it is not raining. Because this happens overhead, few people notice, and many have never seen one. We spend a lot of time looking down at our devices. I guess there is no real reason to look straight up on a regular basis, if you are not a meteorologist. You certainly don't want to make a habit of looking at the sun when it is high. That's dangerous for your eyes.

A circumzenith arc forms in the same way in which a rainbow forms, except that ice crystals rather than water droplets separate light into the 7 colors of the spectrum and create an arc. The ice crystals are in cirrus clouds, miles and miles above the ground, where the temperature can easily be 40 to 50 degrees below zero. If you've seen a solar halo or parhelia, then you've noticed that these and a circumzenith arc all occur in the direction of the sun. A rainbow only happens in the direction of your shadow.

When's the best time to look? Choose a sunny day, when there is a very thin veil of high clouds that does not blot out the sun. Those would be cirrus clouds. We get extra cirrus clouds as a byproduct of airplane vapor trails when planes burn fuel that releases water vapor. You see cirrus clouds any time of year, but the dry, colder air of winter limits the puffy cumulus clouds that would otherwise block a portion of your sky view in the summer. A circumzenith arc may last seconds, minutes or even a couple of hours. It can appear, disappear, and then reappear, all depending upon the clouds. It may be faint or bright or neon electric. A circumzenith arc tells you the clouds are high and cold. If nothing else, it's a pretty sight that typically comes along with calm weather.

A halo around the moon is called a lunar halo. It is easy to spot around a bright moon, but the other accessories that occur around a lunar halo are very difficult to see, because the moon is much less bright than the sun. When there is a layer of middle level clouds of water droplets, rather than ice crystals, around a bright moon, we see a bright fuzzy area of color rather than a ring. This is called a corona. The word corona means something with a glow or spikes around it. The light in a corona is both refracted and scattered.

On your next daytime flight, take a gander downward, at the shadow of the plane. When you fly over middle level clouds made of water droplets, you'll see a similar sight to the corona called a glory. A glory is concentric rings of fuzzy colors of the spectrum centered on the shadow of the aircraft. A glory forms opposite the sun when light is diffracted, refracted, and then reflected inside the individual water droplets of the clouds.

Sometimes when you look up into high clouds, you'll notice random, yet smooth, banding of all the colors of the spectrum. That's called iridescence. Iridescence happens in high, thin clouds made of small and uniform water droplets or ice crystals.

Explore the skies. You might be lucky enough to find a fogbow. That's just like a rainbow except the sunlight is refracted and reflected by the smaller droplets of a bank of fog. It creates an arc but has much less color than a rainbow because the water droplets of clouds are much tinier than raindrops. For a fogbow to form, the sun needs to be relatively low in the sky, at an altitude or angle of 42 degrees or less. Depending on the time of year, formation could be within a few hours after sunrise, or a few hours before sunset. Those are the same times that you can soak in the sight of a rainbow. Similar to a rainbow, a fogbow is found in the direction of your shadow. With a full moon, it is possible to see a faint fogbow and even a faint nighttime rainbow, known as a moonbow.

All of these colorful and unique sights are from light diffracting, reflecting, or refracting in water droplets or ice crystals. Each gives a meteorologist clues as to what's going on in the atmosphere.

Fresh snow is brilliant white. Individual ice crystals are actually colorless, but they reflect all 7 colors of the spectrum to appear white, especially when they blanket the ground. Glaciers often have a blue hue. Similar to air, the longer wavelength colors pass through with little scattering, while blue is scattered the most in glacial ice. Blue color in oceans gives Earth the nickname "Blue Marble." Blue is the dominant color scattered in water, although some bodies of water have different colors based on organisms, minerals, and sediment.

The bending of light doesn't just happen in water droplets and ice crystals. It also happens when the temperature changes fast in a short distance above surfaces. The most common way we see this is in a mirage. Look at the heat rising off of the shingles of a roof and then look beyond the shimmering. A mirage is not your mind playing tricks on you. It's light taking a new path to reach your eyes. Our brains tell us that everything we look at is in a straight line but that's not always how light travels. The illusion of a mirage is something you'll frequently find in deserts. What you expect to see on the horizon is replaced by an inverted and scintillating image of what's above the ground or what's in the sky.

On a hot roadway, we see heat radiating away fast. The huge difference in temperature right above the road surface creates a large difference in air density. Light slows when it passes through more dense air or substances, so the light that seems to reach your eye from the roadway is actually coming from the sky, on a curved path. We see the sky in the road and the shimmering of rising heat makes it appear as a puddle. This is not just something that happens on a hot summer day; it happens in the chill of winter and at night, because roadways absorb a lot of heat compared to the air. When any surface is much warmer than the overlying air, you might see a mirage, or at least scintillation.

Along coastlines, from a distance the trees appear to float above the water. Here again, the light bends due to differences in density of the air just above the water. The intensity of any mirage is controlled by how fast the temperature changes in a short distance above the ground or the water.

Another distortion based on temperature change or air density is when the sun or moon rise or set, they appear flattened. Light from the top of the disk of the sun or moon passes through less atmosphere and refracts less, while traveling faster than light from the bottom of the disk, resulting in a flattened appearance.

All of these phenomena are caused by how light travels with differences in the density and thickness of air, as well as the size and concentration of water droplets, ice crystals, and aerosols. Atmospheric colors help a meteorologist gauge moisture in air, along with cloud height, thickness, and composition. Mirages and scintillation give clues on temperature, air density, wind, and the stability of air. These works of art improve the understanding of the science of meteorology.

There are more things in the sky than there used to be. Airplane vapor trails are numerous and obvious during the day. At night, there are other things that may go unnoticed because we are not looking up as much. Our ancestors of centuries ago were probably more in tune with the night sky than we are. There was less light pollution in communities, people were not distracted or blinded by mobile devices, and without the conveniences afforded by electricity, it may have just been entertaining to gaze at the heavens. Humans have always witnessed meteors and meteor showers and eclipses, and the northern lights, at higher latitudes in the colder months.

Now, there are many more things that happen in and outside of our atmosphere that you'll notice at night. They are all manmade. Spotting the International Space Station is a regular event around much of the world. Every now and then, people will see a weather balloon, or they'll see sky lanterns floating away from some sort of celebration or observance. For people who take the time to watch the International Space Station glide overhead, they often notice other objects gliding across the sky, typically within a few hours after sunset, or within a few hours before sunrise. There are so many things in the night sky which didn't use to be there, based on technology and innovation. More than 8,000 active satellites orbit Earth, with dozens more small satellites regularly launched for communications and internet. The many more rocket launches result in bright lights, streaks and swirls that seem out of place, hundreds of miles away, as water vapor and other gases from rocket exhaust condense to clouds.

There are pieces and fragments of decommissioned satellites, and debris from old rocket missions that also orbit Earth. NASA tries to keep count and says more than 23,000 pieces of "space junk" larger than a softball are orbiting our planet. More than half a million items larger than a marble are out there, too, with over a hundred million smaller objects! Between new rocket launches and returning spacecraft, and space junk that falls into the Earth's atmosphere and burns up, like a failed satellite, we see more stuff, some of which no one can immediately identify.

Alan Sealls

CHAPTER 10

Climate and Climate Change

How do you know when cold air is slow?
When you catch a cold.

Climate is closely related to weather but it's not quite the same. Climate is the average type of weather that you get in a region, mainly based on temperature and precipitation. If you think of weather as the mood of the atmosphere, then climate would be the personality. Our moods frequently change, but our personalities are fairly consistent, although personalities can change over a long time. Climate deals with longer periods in monthly and annual and decadal averages. A climatologist studies average weather over timescales of months, years, decades, and centuries. A meteorologist studies weather over periods of hours, days, and maybe a couple of weeks. Climatology is to meteorology what a marathon is to a sprint. At a glance, a marathon runner and a sprinter look the same, but they approach their competitions differently.

For example, the climate of the Bahamas is warm and humid. That doesn't mean it never sees drought or cold

239

air. In southern California, the climate is dry and mild but that doesn't mean it never rains, even if that's what song lyrics tell you. Climate of a region is controlled by how far from the equator it is; the location's proximity to large bodies of water; the average wind direction; and the elevation.

Colder climates are closer to the Arctic and Antarctic. They result from the lower average amount of sunlight and heat received there. Locations near the poles that have a prevailing wind blowing from open water to land would have wet climates, with much of the precipitation falling as snow.

Warm climates are near the equator where the sun is strong all year long. When the prevailing winds blow from the oceans to land you also find a moist and rainy climate. Wet climates can be found where moist air is forced to rise over mountains, as in the Pacific Northwest. There are many rain forests in that part of the country. A wet climate could be either rainy or snowy or both, depending upon the average temperature.

Dry climates are in parts of the world that are distant from water, or where moisture is blocked by mountain ranges or limited by the average wind direction. The driest climates are deserts. In some deserts, precipitation may not reach the ground in years. The definition of a desert is based only on meager amounts of precipitation, not temperature. That means there are places in the Arctic and Antarctic that are classified as deserts.

Climate can be uniform over regions of Earth but in small areas like islands, you can find multiple climate zones as you travel from the moist, warm coast to tall, cold mountains. When those zones are relatively small, they are known as microclimates. Tall mountains anywhere create microclimates, especially those on coastlines. For any climate, the average or normal temperature and precipitation is calculated for a 30-year period and recalculated every decade.

Air takes its heat and moisture properties from where it sits. A mass of air that resides in a desert would be dry. A mass of air that sits over the ocean near the equator would be warm and moist. Other air masses could be cold and moist if they form over the open oceans near the Poles, or cold and dry if they form over interior land masses near the Poles. An air mass has nearly uniform properties. On a weather map, air masses are regions of high pressure, and they are shown as the letter H. When wind transports air masses and they meet an air mass with differing qualities, weather occurs. Weather, the day-to-day change in the atmosphere, could be wind, clouds, or precipitation.

The meeting zone of air masses is called a front. The front always lies in a valley or trough of low pressure, and it has rising air. When cold air intrudes into warm air, we see a blue cold front, with barbs pointing in the direction of cold air movement. Some cold fronts move through with little fanfare. Others have an obvious signature of stormy weather, followed by a dramatic increase in wind, and decrease in humidity and temperature. No two cold fronts are the same. The impact of a cold front is related to what type of air it is replacing. As a cold front pushes warm air away, it also forces warm air to rise, since cold air is more dense than warm air. That's why we may see billowing cumulus clouds along a cold front. If a cold front is pushing away dry, cool air to replace it with dry, colder air, there may not even be any clouds. When a cold front is moving slowly, especially in early fall, the air behind it may be warmed by the ground, so it might only be slightly cooler. Nights get colder, but days may stay the same or even be warmer after the front passes. That's because dry air cools quickly at night but warms quickly in the day.

The greater the contrast in temperature and humidity from ahead of a cold front to behind it, the more dramatic and sometimes volatile the change in weather may be. The most active and noticeable cold fronts are ones that move fast, plowing into unseasonably warm and humid air, to lift it, push it, and replace it with unseasonably cold and dry air. These are the ones where you would expect thunderstorms that might form a squall line, with a threat of severe weather.

In the heart of winter, cold fronts can get strong as they carry air that originates in the Arctic, to plunge southeastward through the central U.S., driven by a strong jet stream. When that frigid air has a long but swift journey over snow or frozen ground, it doesn't warm up much. This is when deep freezes may extend to low latitudes. Why does all this happen? Earth is trying to balance extreme cold that forms in the Arctic, with warm air always in residence nearer the equator, by mixing the two. Along the mixing zone you find a cold front heading southward, and a warm front heading northward. We don't get to choose.

When warm air intrudes into cold air, a red warm front is drawn with smooth half circles, known as pips, pointing in the direction of motion. If the two air masses meet but neither is moving, then a stationary front is drawn with alternating blue barbs on one side and red pips on the other side. Air masses move daily and weekly, but their average position is a part of climate.

While regional climate zones seem permanent, nothing on Earth is truly permanent. Climate and climate zones do change over centuries. The causes of climate change include differences in the intensity of the sun; differences in the orbit of Earth around the sun; differences in the angle of tilt of Earth's axis; and changes on Earth's surface from volcanoes, earthquakes, and shifts in large bodies of water.

When a large volcano erupts, gases and ash are ejected into the atmosphere. In extreme cases, the gases and particles may create a layer around the globe that can cool the troposphere for months and sometimes years.

Earthquakes or landslides can reroute rivers and drain lakes. They may also dam rivers to create new lakes. Those lakes become a source of moisture for a region.

Human or anthropogenic influences on the Earth change climate. Airplane condensation trails have never been so numerous. As they spread, they create cirrus clouds that would otherwise not have been there. This extra layer of cirrus clouds changes how much heat escapes from parts of Earth, and it also blocks some sunlight from reaching the ground. The amount of sunlight reflected from the entire Earth is known as albedo. A change in albedo means a change in the Earth's energy balance.

Aerosols can alter albedo, and regional temperatures, far from their sources, as they travel and disperse in wind. Smoke particles or soot from industry and transportation can settle on snow and ice to absorb more sunlight. This speeds up melting, resulting in lower snowpack, increased runoff, and warmer air.

Changing climate

The influence of human activity on climate is clear on local levels. Where trees and foliage are removed and large cities grow with solid asphalt, concrete and steel, the temperature is higher. That creates an urban heat island. Large, paved parking lots absorb far more heat than wooded areas or fields of grass. Homes and buildings with air conditioners pump heat outside that would otherwise have remained in the buildings. These small changes in local temperatures can shift average wind. In dense urban areas, the buildings create and modify wind patterns. Combine the warmer climate of cities with vehicle and industrial exhaust emissions, and aerosols, and that can form more clouds, over and downwind of cities, which then transport pollution to outlying towns and may deposit more rain over rural locations. All of this shifts precipitation patterns so that in and around large cities the climate differs from that outside the cities.

Large-scale agriculture also contributes to changes in local climate. Different types of crops emit varying amounts of water while they grow. That's called transpiration. Crops that pull moisture from the ground and add a great deal of moisture to the air can create a wetter climate as they contribute to more clouds and rain. This is significant in arid regions where the water source is mainly aquifer. Farmland that results in fields of standing water leads to increased evaporation and may create more clouds and rain from that moisture.

When a forest is cleared for ranching or development, another shift in average temperature and moisture occurs. The patterns of rain runoff change.

All of these are unintentional byproducts of human life, but they do impact climate locally and sometimes regionally.

On small scales and in small areas, people purposely alter weather. The amount of rain or snow from certain types of clouds can be enhanced by a few percent by adding frozen silver iodide or ice crystals into the clouds. This is called cloud seeding. Cloud seeding is legal in some states and some countries, but the practice can lead to litigation between neighboring jurisdictions when someone claims a negative impact or result.

The largest potential impact of humans on global climate is in the gases we create and release to float into the atmosphere and stay there. The burning of oil, coal, and gas, which are all fossil fuels, releases carbon dioxide. Carbon dioxide is efficient at absorbing heat and essentially trapping it based on its molecular structure. Each year, billions and billions of tons of carbon dioxide are emitted into the global atmosphere from power plants, industry, and from vehicles. If you drive twelve thousand miles each year, in a vehicle that gets 30 miles per gallon, then you are burning 400 gallons of fuel annually. Each gallon of gasoline has about five and a half pounds of carbon, so you directly contribute a ton of carbon to the atmosphere. It changes from liquid to gas but the weight remains the same. That carbon, from a single vehicle, combines with oxygen to create more than 4 tons of carbon dioxide.

Any engine that burns petroleum, gas or coal creates carbon dioxide. Carbon dioxide has a long lifetime in the atmosphere, making it very influential on our climate. Using carbon isotopes, scientists are able to separately measure carbon dioxide emitted from fossil fuel-burning from that of natural sources. While carbon dioxide is a natural part of the atmosphere, and vital to the growth of plants and trees, there's no question that society continues to put many more tons of it into the air each year than ever before. Plants absorb it, the oceans absorb it, but how much and how fast it is absorbed in natural processes does not keep up with the rate of emission. This leads to anthropogenic or human-produced carbon dioxide emissions warming the Earth.

The average temperature of the Earth has risen in recent decades and in the last century. It's not just warmer land where people are. The warming includes land worldwide, along with warmer air globally, and warmer seas covering the planet. While some years may show little change or a slight cooling, the overall trend remains upward. Sea level has increased too. A portion of the sea level rise is due simply to warmer water expanding. The consensus of research scientists is that the use of fossil fuels worldwide is the major cause of a warmer planet.

By itself, a warmer Earth is not a problem. Earth has been both warmer and cooler in the past, without much influence by humans. However, our societies and economies are structured around the current climates. A warming globe is a problem for humanity. Changes in climate result in shifting weather patterns. Wetter areas may become drier, drier regions may become wetter. Warmer locations may become colder while colder areas may warm. Precipitation storm tracks and intensity may shift. There will also be extremes that go to higher levels or increased frequency, whether it is cold, hot, wet, or dry. A warmer planet means expanding oceans and more melting of glaciers leading to rising sea level. Higher seas put low-lying areas, property, and millions and millions of people at a greater threat of tidal flooding and storm surge.

Climate change is a slow process. However, the rate of change in the last century is greater than it's ever been since humans walked the Earth. That's what makes it significant. There are processes in play that we've never experienced. The ones that can make things more extreme are known as positive feedback, but they are not positive for us. For example, if warming results in less overall sea ice in the Arctic, the darker exposed ocean water absorbs more solar radiation and warms further. At the same time, the smaller area of ice reflects less sunlight back to space. These would increase the warming rate, and then the warming would further increase those in a reinforcing feedback cycle.

Along with positive feedback processes is the worrisome scenario of a tipping point. A tipping point is a rapid change, also a point of no return, which unleashes a much larger impact. We can't exactly know climate tipping points because we've never been there before. If you didn't know the melting temperature of ice, just for example, how could you possibly project when it turns to water? It's similar to the proverb of the straw that broke the camel's back. There are tipping points in many Earth processes, and the one that would be catastrophic is if land-based glaciers, like the Greenland ice sheet, melt enough to slide into the ocean.

Supercomputers run complex models of future climate based on what we understand right now. It cannot be predicted with 100% certainty since the current scenario is an experiment we've never conducted before. Unlike a daily weather forecast, which is for a specific condition, at a specific location, at a specific time, where you can verify your accuracy the next day, the verification of climate change takes many decades. However, climate models forecast average conditions, over broad regions, over long timespans. Multiple models, run with different variations, give scientists high certainty of how the climate is changing. Waiting for 100% certainty as to what portion of climate change is related to human activity is time lost. The global warming trend is underway and the sooner we make preparations and adjustments as individuals and societies, the better off our future will be. The largest adjustment starts with reducing the demand and use of fossil fuels.

Oceans have a huge impact on climate and weather with their currents and salinity and ability to absorb carbon dioxide. Increased CO_2 in oceans gradually makes seawater more acidic, lowering the pH balance, which could negatively impact marine life. Weather and climate impact the oceans too. Prevailing wind patterns create ocean currents. Short-term winds create mixing of ocean water. Oceans are directly linked to the air and wind, but in ways that meteorologists and oceanographers don't fully understand. The power of water to hold heat and moderate the Earth's temperature cannot be overstated. Just the top layer of the ocean retains more heat than all the air on the planet. So much of the recent global warming has gone into the oceans. We may not see or sense that in our daily lives, but the impacts will be widespread and long-lasting.

As climate changes, there are immediate losers but there are also immediate winners. The winners gain a more favorable climate for agriculture, recreation, personal comfort, and transportation. Shipping lanes in high latitudes may have less ice in winter to enable more and safer commerce. The losers immediately and in the long-term deal with loss of arable farmland, or mass migration of populations to more hospitable areas. There may be increases in airborne infectious diseases or increases in insect pests. With climate change, some might have to contend with loss of property along coastlines, more extremes in temperature, storms, precipitation or drought, and likely civil unrest and political conflict over dwindling water and land resources. The key word in climate change is change. Some change can be prevented through proactive action. Not all change will be dramatic or immediately negative, but all change does require adapting and adjusting. For humanity, the resulting changes are more negative than positive.

Weather extremes create stresses on society, but no one storm or weather event can be totally blamed on, or credited to, climate change. Climate change shifts the odds or frequency that a particular storm or weather event will happen. The impact of any weather event increases with population growth and density, even if the weather event is of the same magnitude as a similar event from decades ago. Traditional and social media, streaming video, live web cameras, traffic cameras, doorbell cameras, and now drones give so much more continuous information, often live, of things we may never had seen or heard of long ago. This creates a difficulty for individuals in comparing weather impacts over time on a warming Earth.

As individuals, especially those of us who are seasoned in life, we tend to make comparisons to past climate and weather patterns in our younger days. Be careful with that because our bodies and physiologies, hearing, taste, vision, and perspectives change dramatically over many decades. Years ago, my oldest brother called from Boston and said, "It's never been this hot before, I broke down and bought and air conditioner." I said to him, "you've never been this old before!" When I was a kid, I remember the snow being over my knee, but back then, my knee was a lot lower to the ground! Our occupations and hobbies skew our perspective of weather events. Those who work indoors and travel in climate-controlled vehicles have a different sense of the elements than those who work outside. We become acclimatized to certain conditions with time. Local changes in communities caused by development can't be assumed to extend to the rest of the world as a global change. Even for those of us who grew up, lived, and worked in the same city for all of our lives, that still is not a proxy for the remaining planet. Similarly, lack of local change does not extend to the rest of the globe either. This is why scientists use global and historical data to arrive at objective assessments.

Carbon dioxide is the main gas behind a warming world climate but there are other gases like methane and nitrous oxide that absorb and trap heat too. Methane is also potent.

On a warming planet, much of the warming actually comes from higher evaporation rates, increasing water vapor levels in the air. Water vapor is an abundant natural gas highly capable of retaining heat. It keeps the Earth at a temperature conducive for life. While human activity is contributing to warming the planet, we are unable to fully separate that from natural cycles. That does not mean a warming planet should be ignored. There are ways to estimate how much of a change from regular cycles is ongoing. A climatologist can look at and reconstruct past climates using ice cores from glaciers, sediment cores from the bottom of lakes, historic diaries, writings, and sailing ship weather logs, and by studying old trees.

The rate of growth of trees depends greatly upon the temperature and rainfall. Each year, trees add a layer of growth which appears as a ring, seen when you cut through a tree trunk. The wider the ring spacing is, the more likely that rain and temperature were above average in a given year. This allows estimates of climate going back several hundred years from living trees. Recovered or petrified trees from centuries ago shed clues going much farther back in time.

In the bottom of lakes, sediment and biological debris settle in layers, year after year. By drilling down into the mud of lakes to recover what's called a sediment core, a climatologist can determine what type of plants and insects were dominant in the past, which relates directly to climate. Sediment layer thickness tells if large storms occurred to deposit unusually large amounts of sediment in specific years.

Ice cores give historic information through trapped bubbles of gas. These gas bubbles are a snapshot of the atmosphere centuries ago, allowing us to know the actual type and ratio of gases. Ice cores from glaciers can easily be tens of thousands of years old. A glacier is ice from compacted snow that never melts. It may be hundreds or thousands of feet thick. Glaciers form over centuries in the coldest climates and in the highest elevations of tall mountains. Glaciers are essentially rivers of ice that move downhill so slowly that your eye cannot see the motion. When they reach water, the edges break off, bit by bit. That's called calving. The ice of glaciers often has a blue tint and contains freshwater.

Some glaciers are shrinking as a sign that the planet is warming. Glaciers or large areas of ice on land that melt will raise sea level. Ice that melts, already in water, won't change sea level. Glacier and ice melt can send tremendous amounts of very cold water into oceans, where it impacts ocean circulation because cold water is very dense.

In parts of the world, some glaciers are actually growing. A warmer atmosphere holds more moisture that will fall as heavier snow in cold regions, since the temperature near the Poles and at high elevations will still be below freezing, just not as far below as it once was.

One of the big misconceptions of a warming planet is that every place gets warmer. Some regions will cool down, as the jet stream may slow and take more of a north-south route, more frequently. A slower jet stream on a new path can cause longer periods of either wet or dry or hot or cold weather. It could also result in wider weather extremes in any given location, over the course of years. Coastal communities may also cool down if the ocean circulation carries colder water to that part of the planet. Those same areas that get colder ocean water from melted glaciers might see sea level fall since cold water is more dense than warm water. Given that the northern hemisphere has more land than the southern hemisphere, changes around the planet won't be uniform.

Weather extremes in a single city or county should not be expected to mimic or dictate changes for the planet. Even a large state is only a fraction of the Earth's surface. Consider that the continent of North America is only 5% of the surface of the Earth. The United States is much less than that. The majority of the planet is covered by water. Your personal experience of shifting climate or steady climate cannot possibly give you an idea of large-scale trends over the globe. For that, we must look at reliable data and records.

A warming Earth means the likelihood that drought, flood, heat and even cold will shift, region by region. It's crucial to know that some locations may see no significant impact from climate change. As climate changes, local weather will always deliver a variety of conditions and there will still be cold, intense snowstorms, and record cold and snow. As no single weather parameter controls tropical cyclones or tornadoes, no one can say for certain how those might change in a warmer world. Warmer air and higher moisture in air would certainly make more fuel available for storms, but wind and air pressure are other factors in storm formation that create the mysteries. In all climate and weather, change is constant and inevitable, but an aware, flexible, and proactive society handles the changes more smoothly.

Climate science, reason, and belief

Meteorology and climatology both are science, not belief. Science can be proven through measurement, experiment, and demonstration. What's odd about science is there's so much that happens that just does not make sense. How is it that the Earth's orbit around the sun is precise and continuous? How can an acorn become a mighty oak tree that lives for centuries? How can tons of rain fall from a cloud, something that floats? Why is it that gravity holds us to the planet? For what reason does electricity fly through the sky as lightning? If you take away the principles of science, none of this seems reasonable. Go from logic to observation and experiment, to find the reasons why.

When I tell people that there was a recent record high, no one questions how I arrived at that conclusion (thank you). People trust that my training and track record show I kind of know what I'm talking about. So why is it that a small but loud group of people question and debate climate researchers? Survey data by George Mason University shows attitudes toward climate change largely align with political leaning. Whenever I see an online article about climate change, I hold my breath as I skim the comments. Without fail, comments include, "The Earth's climate has always changed," dismissing the decades of study that thousands of climatologists around the world put into learning it. From there, comments become political and accusatory.

Have you ever met a climatologist? I have. I've worked on boards and committees with them and interacted socially. While they may not be as cool and charming as meteorologists (ha ha), they are passionate, intelligent, human beings who dedicate their energy to finding fact. They didn't go to graduate school and labor to earn PhDs to seek a position where they make up stories and falsify data to mislead and make money. Climatologists deserve respect.

When your physician tells you she sees long-term changes in your labs that need your attention, do you dismiss that professional perspective with, "They've changed before"? When your truck goes in for service and your mechanic says, "Your engine diagnostics continue to show a problem," do you dismiss that with, "I saw a post that says mechanics make things up to get your money"? Then, why do people do that for climate change?!!!!

The basics of climate change

Science is neither opinion nor emotion, nor number of followers on social media. To fully understand climate and weather you need to have a background in physics, chemistry, calculus, thermodynamics, and fluid dynamics. In many ways, both studies are complicated but in other ways they are simple. The sun heats the Earth, more at the equator than at the poles. The Earth rotates on a tilted axis, orbiting the sun, to create seasons. Earth has land, water, and air which all respond differently to heat. Those factors create weather, as the Earth constantly uses air to try to balance heat and moisture. Climate is the average of weather over long periods. Weather is what you dress for. Climate is what's in your closet.

Everything on Earth has always changed, often on time scales that far exceed our lifetimes and written history. Meteors strike. Volcanoes erupt. Continents drift. Lakes and rivers form and disappear. Glaciers grow and melt. Species are born and die. Some football teams actually win a game or two.

Global climate change is hard to grasp because most of us travel in relatively small circles. While air keeps our planet at a habitable temperature, carbon dioxide ($CO2$) and methane ($CH4$) are big reasons behind a warmer Earth. They are known as greenhouse gases because they trap heat similar to the way a greenhouse holds heat. Methane is burned for energy, and it's a byproduct of some industry. It is emitted from livestock. It also is released along with $CO2$ as some permafrost melts. Melting permafrost can release bacteria and viruses that humans have never encountered.

So, what's the problem with using fossil fuels that emit $CO2$ and methane? $CO2$ is what plants live on, and warm air helps them grow. These are good things, right? Like vitamins? Well, yes and no. There's a point where increasing any nutrient creates a negative outcome. You would never triple a vitamin allowance or a medical prescription, thinking that you will get healthier three times faster. That would be dangerous to your health.

Plants can't take in all the excess CO2 that we create. The oceans absorb a great deal but that's bad for corals; and as the oceans warm, they take in less. Large amounts of CO2 trap heat within the atmosphere and THEN, more water is evaporated into the air. It's the extra water vapor in air that amplifies the warming of the planet.

So, how do we know that this cycle of warming mostly results from human activity? Thousands of peer-reviewed research studies, using physics, chemistry, calculus, thermodynamics, and fluid dynamics applied to the properties of gases removes doubt. Computer models recreate past climate and project future climates, allowing scientists to add or remove factors like CO2 to see different solutions. This includes using historical data to predict what scenarios could have gotten us to the current state of Earth's climate. Like daily weather forecast models, they are not perfect to a pixel, but they are hugely in the range of reality.

Who is to blame for climate change?

Peer-reviewed research studies and data from the overwhelming majority of climate scientists worldwide attribute human activity as the primary cause of the rapid temperature rise. There's little debate on that. When there's debate about climate change, it usually involves a non-scientist over what we should or shouldn't do about it.

Satellites pointing at the sun confirm that the sun has not substantially changed output in decades.

Since the industrial revolution, increasing fossil fuel use to provide energy to produce goods we need and the quality of life and convenience we want has allowed societies to prosper. And there are many more of us on Earth, demanding more stuff. There are also many of us individuals and companies who want to pay as little as possible for products and services, without considering the true cost of those products. Blame is placed on other countries for being greenhouse gas emitters but those are the same countries that give us a portion of what we want, and the low prices we demand.

Warming from methane and CO_2 production is slowly throwing off the balance that our societies have been built on. Rising sea level is an immediate threat. CO_2 and methane are potent. There are many tiny things that can have huge negative impact—snake venom, fentanyl, a tick, coronavirus. How can we insignificant humans possibly influence the planet? Think of termites. One termite is no problem. Thousands are a problem. Our individual footprints on Earth are small but as a society they are tremendous. We've dammed and rerouted rivers, trimmed the tops off of mountains, filled wetlands, removed groundwater, and introduced invasive species.

A change in climate means changes in weather patterns, although I don't know exactly how our weather may change. What do we do now?

What to do about climate change

Earth doesn't care if it warms. We care because we don't like change. A change in weather and water patterns will shift farming and food production. Agriculture will adjust, and we'll all pay for that in the grocery store. In poorer countries, changes may lead to mass migration and civil conflict where resources are scarce. Shifting climate zones will change the range of wildlife, pests, and insect-borne diseases. All of that is a direct threat to human health. Our physical world is highly interconnected. Coronavirus is a hard lesson in interconnectedness.

I don't have a slam-dunk answer as to what to do about climate change, but I do have suggestions. We can't flip a switch to stop what is already in motion. Some of the ideas I've heard about sending things into space to reflect sunlight sound like the beginning of a horror movie. Which countries would have the legal and moral authority to do something to the Earth? Who would pay for it? What if it were to go wrong?

Climate change is like having joint issues develop because of your dedication to an extreme sport that your doctor told you to step away from. Joint problems do happen naturally, but lifestyle can be a contributing factor. You're then faced with having to adapt to living with your condition or seeking a magic pill or procedure to fix it. There are no magic pills or procedures that come without side effects, cost, and consequences.

Is it hopeless? No. You have the power to live with the Earth. Adapt, conserve, reuse and recycle. In all of your lifestyle and purchase decisions ask yourself if it's good for the planet or if it requires large amounts of energy from fossil fuels to create and maintain. You are a part of the solution with your wallet and your ballot. Vote for elected leaders who understand the potential negative impact of climate change on your community and region.

We learned in the 1980s that man-made chemicals, like the propellants in spray cans, were destroying beneficial ozone in the stratosphere. Countries acted to ban those chemicals, and data shows the ozone levels are recovering. We have power.

Think of the billions of creatures around the globe that do not compete with each other to acquire and hold stuff, as humans do. They limit their footprints to what's needed, not what's desired based on social pressure and expectations. Birds build nests of organic material (with scraps of non-biodegradable waste from our lives). Bees take pollen from plants but in the process, they help the plants to propagate. When dogs poop, they try to bury it (except for some of the dogs that walk past my house).

Adapt to climate change, as individuals and countries. Limit contribution to future change by conserving energy. That puts less fossil fuel in the pipeline, literally. Reuse, repair, donate, and share stuff to give it a second and third life, demanding less energy usage for further production. Recycle, to also reduce energy production and limit the need for more raw materials. Don't waste food.

Consumer demand drives production and transportation, which remain very dependent on fossil fuels. I can't think of any industry that exists just to produce greenhouse gases. They exist to provide goods and services at the lowest short-term cost to you and me and others who want to pay the lowest price.

Among the obvious task of reducing fossil fuel use, is the need to sequester CO_2 that's being created, and to capture and reuse or repurpose CO_2 that is in air. That's known as direct air capture or DAC.

There are alternative and greener energy sources but developing and implementing them has cost and a carbon footprint. They also have job and investment opportunity. I don't have the expertise to say which is best or most efficient. Support businesses that limit planet-warming practices. Through your choices and actions, you have the power to make positive change, especially when you follow accepted principles from reliable and trusted experts. Change is inevitable. Jobs and industries must pivot. I would bet that if you went back in time, blacksmiths were upset when the automobile was invented because it meant they would make fewer horseshoes. Think about candlestick makers being concerned about the widespread use of electricity. Those are two occupations that had to change for the world to move forward to one that was more productive, more safe, and more efficient.

For the people who immediately talk about the politics of climate change, rather than the simple science, don't take the short view of resisting change when the long view shows that managing change can create resilience for communities, better health for individuals, and financial opportunity for businesses and for people.

Science trust

Don't get caught in the rhetoric of, "You can't trust science anymore." That is a misguided, canned phrase, which goes against common sense and reality. The first person we met, when we entered the world, was likely a scientist. A physician or nurse trained in medical science made our transition easier and safe. You probably don't remember.

No matter what negatives are uttered about science, people do trust it, even if they don't want to. Science is not a belief, and it's not an option for living. When you used the commode this morning, you had 100% faith that when you pushed the lever, the stuff in the bowl would go down, rather than shoot up. You knew the science behind the engineering, and the principles of water pressure and gravity would work for you, even if you don't know how they work. When they don't work, yuck!

Whether you had a glass of water, or bottled water today, you didn't question the purity and safety of it. Thanks to science and chemistry for that, and federal regulations that put your health first.

The kitchen chemist knows that fundamentals of science allow the preparation of meals that are safe to eat. Break those rules and you might be visiting the commode again, or an emergency room.

As you read this, if you are holding a paper copy, you probably are reflecting on the biological science of growing trees for harvest, and then creating pulp through a chemical and mechanical process that results in paper. Others of you are saying, "Paper! Who reads anything on paper anymore?" Good point. You, instead, are relying on the transfer of electrons to power your device, and diodes to illuminate your screen. There's that sneaky science again, making our lives easier.

In any form, reading is possible due to the electromagnetic spectrum and the science of sight. For those who might listen to written text through text-to-speech software on a device, there's definitely science behind that software, along with the fact that all sound is the physics of vibrating air.

I am absolutely sure that while some people say science shouldn't be trusted, they do trust it. Otherwise, they would never drive a car, for the fear that applying the brakes wouldn't follow the laws of physics.

With regard to greenhouse gases, it's more and more obvious that if nations reduce greenhouse gas emissions, the gases that are already in air will linger for decades, continuing to warm the planet. There will always be some amount of human-produced greenhouse gas, so now we look at technologies to capture and reuse or store the greenhouse gases that we produce. Once again, another large financial opportunity for those who can rise to the challenge of developing and perfecting capture and storage technologies.

Weather modification

"I can turn the gray sky blue. I can make it rain, whenever I want it to." The Temptations were on to something when they sang, "I can't get next to you." That part of the song was about controlling the weather, or weather modification. Weather modification is science, but if you think weather modification only means controlling weather at will over large areas, that is fantasy. I thought of that when I saw a social media headline about people blaming weather modification for flooding on another continent. I don't know where it was because I didn't read the article, because, well, it was social media, and that's not my first source of science information.

There are scary and unsubstantiated claims that have circulated the internet for decades that somebody or some government is controlling the weather. These often fall under the blanket of geo-engineering. These claims take a little bit of science and history and mix it with a lot of myth, misunderstanding, speculation and sometimes lies. For example, the fact that we have so many high-flying commercial jets now results in vapor trails that lead to more cirrus clouds through the natural process of condensation, although some people claim this is a worldwide conspiracy to release dangerous chemicals.

Changing any of the three ingredients for weather—heat, air, or moisture—will change weather. Weather modification can be as simple as using downwash from helicopter rotors to mix air and dissipate fog on a runway. It may be the use of heaters in a citrus grove on a winter night to keep the temperature just above freezing. It may be the inadvertent byproduct of industrial areas of heat along with smoke particles to increase local temperature and increase clouds and rain downwind. Large farming regions modify weather. The United States Geological Survey says that in growing season, an acre of corn gives off more than 3,000 gallons of water daily through transpiration. What about snow machines at ski resorts? The snow ends up coating the ground as a blanket of white, to reflect sunlight and insulate the ground. Large dams create lakes that greatly increase local evaporation. So much of the geo-engineering we do to make life comfortable unintentionally creates small modifications to weather. Climate change is an inadvertent large-scale weather modification.

There are many documented projects and endeavors by governments, military and by scientists around the world to see how much we can purposely control or modify weather. The results show limited, if any, success, and that success typically is only on a local scale. There is a legitimate and legal process called cloud seeding, whereby dry ice or frozen silver iodide particles are dropped into rain clouds from an airplane to increase the rain yield. The main purpose is to boost agriculture. It's not like you can make rain from nothing. There must be rain clouds already, and that makes it really difficult to prove that enhanced rain was due to the seeding. Within the cloud seeding industry, an increase in rainfall of 10% pays dividends.

There is a federal requirement that any non-Federal entity performing weather experiments or operations that might impact a wide area must register with the Department of Commerce. That's in part to dissuade dangerous or harmful undertakings by evil masterminds, or your neighbor. So, can you make it rain? Yes, with a garden hose.

Project Stormfury

"Project Stormfury" sounds like a movie you stumble upon while late-night TV surfing, about a villain with a bad childhood, conspiring to control global weather with laser beams, unless his demands are met. That's not what it is or was. Project Stormfury was a project, described by the Hurricane Research Division of NOAA's Atlantic Oceanographic and Meteorological Laboratory, as "an ambitious experimental program of research on hurricane modification carried out between 1962 and 1983." It was an attempt to see if we could weaken hurricanes.

Project Stormfury was no secret, it was filed for approval with the Weather Modification Board of NOAA, in the U.S. Department of Commerce. It shows up in the Federal Register multiple times in the 1970s as an experiment "to test the hypothesis that the maximum winds of hurricanes can be reduced by at least 10 to 15 percent by seeding the proper clouds with freezing nuclei (silver iodide)." The idea of cloud seeding is to take clouds that are growing and add particles to them to allow water to condense and precipitate faster, to limit the growth, limit hail, and often to increase rainfall. In the case of a hurricane, this would theoretically keep the core weaker and more spread out, and supposedly cause it to rain itself out. Frozen silver iodide has an impact similar to ice, in inducing raindrop formation.

In the 1980s, the project ended. The actual experiments were limited due to few approved hurricanes to test, and due to limited capability of the aircraft used. Long-term results showed no substantial difference between hurricanes that had been seeded, and those that had not been seeded.

Stormfury followed the earliest documented government hurricane modification research project from 1947 in which crushed dry ice was dropped into hurricanes. That was known as Project Cirrus. Results were non-conclusive, and that project actually was perceived to have caused a hurricane to turn, strengthen and strike land. We know now that the path of a hurricane is not something we have any control over. The change in the hurricane path and strength was a coincidence.

Since Stormfury, there have been numerous other research campaigns to see if humans could intentionally increase or decrease rainfall, mitigate hail, and reduce wind. The theories of atmospheric physics and dynamics say it is possible to a small degree, over a limited area, but 100% proof that we did or didn't make something happen that wasn't or was going to happen anyway is near impossible.

When you consider that a single thunderstorm has more power than a nuclear bomb, it makes it easier to understand why we can't manipulate massive storms like hurricanes, composed of thousands of thunderstorms. For Category 5 Hurricane Andrew, striking south Florida in 1992, NOAA's Hurricane Research Division says that the heat energy released around the eye was 5,000 times the combined heat and electrical power generation of the Turkey Point nuclear power plant. That's tremendous power.

CHAPTER 11

Winter

Where does a snowman get his money?
From the snowbank.

W e hit the reset button every January on the calendar, but not for the atmosphere. The start of a new year brings new numbers and statistics for weather and climate. It's purely a mathematical change, almost like a fiscal year, since there's got to be a way to categorize and compare past weather with current conditions and future projections. Even though winter, from an astronomical definition, starts in December, the coldest days are typically in January and February. It's a lesson that Ray learned …

Ray stepped off the bus and turned his collar up. A sharp wind caught him from the right and almost took his hat away. He hunkered down and slapped his palm to his hat. "One of these days I'm actually going to listen to what they say in a forecast," Ray thought. "That's got to be one of my New Year's resolutions." As in the past, Ray was caught off guard, underdressed for the season. It was no problem earlier in the day when sunlight flooded the north-south urban streets and winds

273

were light. Now late afternoon, the view of the golden globe in the sky was replaced by dull gray sheets of clouds and brisk winds. Ray hurried to get under the shelter to wait for the next bus. He watched pieces of litter race down the sidewalk and end up in little swirls in the intersection.

Ray dug his hands deeper into his pockets and turned his face away from the wind. "Are you done with that?" Ray asked the woman next to him, who had just closed her newspaper. She looked at him coldly, paused, and said, "Here." She thrust the paper at Ray and looked away. Ray thought, "I only asked for the paper, not attitude. It's cold enough out here." It was a slow news day. The headline read, "The days are getting longer!" The article pointed out that the minutes of daylight are increasing now that we've passed the winter solstice of late December. Buried in the third paragraph, there was one line that caught Ray's attention. "Our coldest winter days are yet to come."

"Well, what's this all about? That doesn't make sense," Ray said aloud. The woman who had given him the paper looked at him and stared. "Did you read this?" Ray asked her. She paused, looked away, exhaled a deep breath, and replied, "Read what?" "The story on the weather." "Oh. Yes. I was just talking to one of my students about that today," the woman stated. "You're a teacher?" Ray queried. "Yes, for many years." She took off her gloves and reached into her bag. Pulling out a notepad, she started to draw a diagram. "Here's what's happening ..."

Ray looked at her sketch of the sun and the Earth. "I seem to recall an answer from "Jeopardy" that says the Earth is closer to the sun now than it will be in July. How can we still get so cold?" "I give you an 'A' for your brilliance and the sun gets a 'C' for its lack of the same," the woman quipped. "Compare the winter sun to the summer sun in the Midwest. We see the sun in January for around nine and a half hours each day. In the summer, the sun hangs around much longer, up to 15 hours a day. The sun is our heat source. Just as a pot of vegetables left on the stove for a longer time gets hotter, so does our temperature under the summer sun." "But when you turn the heat down low ..." Ray added. "Exactly! Shorter days mean less time to heat up. But there's another difference to the sun in the winter compared to the sun in summer. January's midday sun is very low in the sky. Our shadows are long, and the sunlight is spread out over a larger area. The January sun's rays pass through more atmosphere and weaken before they get to us. The summer sun rises higher in the sky, giving us a more intense beam of light."

Ray looked toward the curb. "I get it! You're saying that until the days get much longer, we can't expect much warmth." His bus was approaching but the woman continued, "That's right. What's really interesting is what you noted; our planet orbits the sun in an egg-shaped path so that we are closest to the sun in January and farthest in July. The change in distance is rather small and therefore has nothing to do with our seasons or temperature changes." Making a motion for the bus, Ray fished his transfer out of his pocket. "I don't want to be rude but ...

Thanks for the paper and information." "Listen, you'd better get some warmer clothes on out here. It's January. Happy New Year, sir." The woman winked and smiled.

You might remember from Earth science class, that even though winter is the coldest season for the northern hemisphere, it's actually the time of year when Earth is closest to the sun, our nearest star. We are only 91 million miles away from it in early January. That's called perihelion. In early July, we are over 94 million miles away. That's aphelion. The Earth's orbit around the sun is a little lopsided. Strange, but true. Impress your friends with that astronomy trivia which tells us that it's the sun angle and length of day that control the seasons, not the changing distance between the Earth and sun.

Wind chills

Winter cold is more intense when the wind blows. That same movement of air that we long for on a hot summer afternoon is an unwanted visitor in the cold heart of winter. There's a big difference waiting at a bus stop when it's 35 degrees and calm, versus 35 degrees and windy. The wind chills you and me. Wind chill is how your body feels when the air is cold, and the wind blows. It's based on research on how fast our skin loses heat, as wind carries it away. When our bodies can't generate enough heat to keep our overall temperature near normal, skin and body temperature begins to fall. That's the chill brought on by the breath of wind on your neck, or your hands, or your face, or on your body if you are underdressed.

Wind chill, also called wind chill index or wind chill factor, comes from a formula. Search online for a wind chill calculator or wind chill chart to get the value. There are many formulae that produce the number but the end result of most of them is in the same range. There has a been a wind chill since our earliest ancestors roamed the planet, but it was not put into an accepted formula until the 1940s. It became more used in weather broadcasts and reports in the last half century, as we've learned the dangers of frostbite and hypothermia. Just like running a fever can be bad for you, if your body temperature falls by a few degrees, that can be bad too.

Wind chill is a big deal. If it's 40 degrees, with a 15-mph wind, your skin loses heat to make you feel like it's 32 degrees. If it were 30 degrees with that same 15 mph wind, you would feel like it's 19 degrees. Frostbite and hypothermia would happen more quickly. If your skin is wet from rain, snow, or sweat, there's an even faster loss of body heat from evaporation. Avoid wind chill on a windy, cold day simply by limiting the amount of skin exposed to the air and by not allowing your clothing or skin to stay damp or wet.

Can a temperature above freezing with a wind chill below freezing freeze your plumbing, or the water in your dog's bowl? No. Wind chill, used in weather forecasts, only applies to exposed human flesh. That means your cat and dog feel a chill but it's not the same chill you feel, because they have fur coats. I'm sure if they could talk, they would convince you on a cold day that they need to be inside by a warm plate of food. They would probably say that on any day.

While air temperature is slow to change, if the winds gust, the wind chill falls quickly, and then rises when the gust dies down. It fluctuates with wind gusts. Wind removes heat from any object fast, but wind chill cannot lower the temperature of an inanimate object below the actual air temperature.

Cold fashion

Dress for the season, the region, and how well your body is acclimated to where you are. Based on physiology, age, and prescription medications, our bodies feel heat or cold differently, so I can only address the meteorology.

In mid-January, in Honolulu, Hawaii, a typical nighttime low is in the middle 60s and a record low is in the middle 50s. Aloha! In Fargo, North Dakota, a typical nighttime low is near zero. Yes, zero Fahrenheit! Record lows for Fargo are anywhere from 30 below zero to 40 below zero.

A visitor from Fargo, visiting New Orleans, would delight to have nights in the 30s, given that January days in Fargo average close to 20 degrees. Your cousin from Honolulu would shiver and quiver with nights in the 30s. From different regions, we acclimate to certain temperature ranges. Our adjustment varies by season, thanks to the overflowing closets full of options. However, we also adjust to long periods of the same temperature range. That means that if we have 5 winter nights in a row that are milder than average, when they actually are average on the 6th night, it will feel cold.

As you travel and hear a forecast for "cold" from a meteorologist, it is relative to average temperatures for the location and season. Calling a winter night "cold" gives you an idea of how much effort and money you have to put into heating your home. Those with newer homes and well-insulated homes may not be concerned, but for the many who are in older homes, with less insulation, "cold" has a different threshold and it is a big deal. Even if it's not cold to you, keep warm thoughts for those who may not be able to find warmth.

Winter is the season that challenges our senses. It requires preparation for the snow and cold. However, if you're like I am, then you don't really like to wear hats or boots. As mom always says, "Wear a hat." Even when it's not very cold your body loses a lot of heat from your head (some of us lose more than others, based on head size!). A hat allows you to retain heat and it also keeps your head dry when precipitation occurs.

Gloves are a necessity, but mittens are even better. As goofy as you may feel at first wearing mittens, you'll appreciate the fact that mittens allow your fingers to share the heat from one to another. I admit it is hard to handle a smartphone with mittens, but it would be even harder with frostbite. Frostbite is more than numbness of your fingertips, toes, and face. It's when your flesh freezes. If skin freezes and loses color and sense of feeling, you are approaching danger. In minor frostbite, the effects are similar to a third-degree burn. Medical treatment would be wise. Severe frostbite of fingers or limbs requires amputation.

Earmuffs and scarves—yes. Boots—yes. Layers of clothing—double yes. The extra gear might seem inconvenient but don't leave home without it! Let your common sense override your fashion sense. Cover exposed skin. Avoid having rain or snow soak your clothing. When you walk around in damp clothing you can count on being cold. Dress smart and follow the data that is presented in weather forecasts.

"It was so cold, that ..." What is it that makes people exaggerate how extreme the weather is? While TV newscasters are often charged with hyping the weather, just listen to those around you. In most cases, people knowingly embellish, with no harm done. After a big snowfall, I've heard more than one close relative of mine say we had a blizzard. Wrong! Calling heavy snow a blizzard is like calling potato chips a meal. They are not really the same. Now that I've whet your weather appetite, let's look at our winter menu. We'll browse the words and weather served to us in the cold season.

"Winter Weather Advisory." Look for a mixture of snow, sleet, or freezing rain causing slower driving and walking conditions. This advisory will have different criteria in different parts of the country, based on the climate, and how prepared communities are for the hazards.

"Wind Chill Advisory." Low temperatures accompanied by strong winds, resulting in your body feeling colder than the thermometer indicates. Spending long periods in the cold and wind are hazardous.

"Winter Storm." Heavy snow, which may blow and drift due to high winds. Traffic will be impacted, especially on untreated or unplowed roadways. Also possible are rain, sleet, or freezing rain.

"Blizzard." Steady winds above 35 mph causing blowing snow, with minimal visibility, regardless of temperature. True blizzards are not as frequent as many people might believe. A blizzard can occur on a sunny day, with no clouds, when high wind picks up snow and blows it around to create a whiteout.

Cold air flowing across large unfrozen lakes, like the Great Lakes, picks up vast quantities of moisture and heat to form clouds over the lakes. The clouds move with the wind, often in bands or streets, to release snow downwind of the lakes. This is called lake effect snow. The falling snow can be very heavy and produce local winter storms, even with otherwise quiet weather and nearby clear skies. In a typical winter, long periods of extreme cold will eventually freeze portions of the Great Lakes and other high latitude lakes.

Winter above the Arctic Circle has sea ice growing. Given the importance of shipping and commerce, the U.S. Coast Guard works to keep shipping lanes open in winter. While ice is less dense than water and therefore floats, thick sheets of ice can stop or strand a ship.

In the clear aftermath of a winter storm and lake effect snow, satellites show extensive areas of land blanketed by freshly fallen flakes of snow. One way to tell the snow apart from clouds is that the clouds cover everything, but snow reveals lakes, rivers, coastlines, and other geographic features. On the ground, after a fresh snow, the snow absorbs sound to result in a hushed environment. On the other hand, when extreme cold air sits in place, with little wind, sound can travel farther, so we often hear distant things on calm winter mornings that we don't hear on windy or warmer days.

Cold stress

Dry air is a winter inconvenience. I find my skin to be much drier in the winter. Cold air quickly evaporates moisture from our skin because cold air holds little moisture. To conserve your body's moisture, use creams, lotions, or oils. It's not just a cosmetic concern. Those of us with very dry skin can suffer flaking or cracking and that becomes a health issue. To replace lost moisture, drink lots of water. Aside from dry skin, some folks get dry throats, eyes, and nasal passages. In extreme cases, a person might need medication to alleviate the dryness.

Modern technology is part of the dry winter air problem. In many homes and buildings, the heating systems suck in outside air. When cold air is heated, the relative humidity of the air goes down, way down. That's not good for our health. The solution to dry indoor air is to add moisture to it with a humidifier. Newer heating systems accept humidifiers which add moisture to your air. Older systems such as radiators simply have an open container that you fill up every few days with water. If you have neither of these then all you need to do is keep water in a container close to the radiator where the heat will evaporate it into the air. Notice how fast the water disappears when it's bitterly cold outside. Adding moisture to dry winter air will somewhat soothe your body's dryness. WARNING: Do not keep water near electric heating systems!

Another winter stress is the concern for freezing pipes in homes. Water expands when it freezes and that can burst a pipe. The risk varies by region and by home construction type and age. It's more of a concern in southern regions of the U.S., where freezes are less frequent. Let it drip. That's the advice I give to anyone who asks me if they should let the water trickle in their plumbing to keep pipes from freezing. It's a simple answer to a complicated question. It's complicated because there is often not a yes/no answer, or at least one that I can give!

With a forecast low temperature, just for example, of 30 degrees, water will freeze, but there are a bunch of things you have to consider if you are concerned about your plumbing. A forecast is an average for an area, not for your personal property. Even if it were specific to your home, depending on the size of your lot, you can easily see a three-degree temperature spread, where parts of your property may stay above freezing. If you live on a hill or slope, it will often be colder at the bottom of the hill than at the top. If your home is fully exposed to wind, that would mix the air and might keep you a few degrees less-cold than on a totally calm night. If you are surrounded by large trees, still with leaves, those will help to keep you from being as cold as you would be with no trees. If you are right next to a coastline, a bay, or a large lake, the water will moderate the air temperature, especially when a light wind blows across the water to you. Similarly, if you live downwind of an industrial or shopping complex, the heat from that may help you too. Those are environmental influences. Now, what about the things you control?

If your home is raised above the ground or has a crawl space, then your pipes are more at risk to freeze. Are your pipes metal or PVC? I don't know! Is your plumbing properly insulated? I have no clue. Was your plumbing built for cold? Don't ask me. These are things that only a plumber or builder can answer.

Nobody can guarantee you won't have freezing pipe issues when you take the environmental differences of where and how a home is situated, and the physical differences of how homes are built, and how far below freezing the temperature will get, and for how long it will be below freezing. If your neighbor's home is similar to yours, ask them how cold they go before letting water drip, or at what temperature they've had plumbing problems. When it's really nippy, some need to get drippy.

Polar vortex bomb cyclone

Have you seen winter headlines with catchy phrases like "horizontal hurricane" or "atmospheric river"? Those are about moisture that streams into the west coast of the U.S. and creates flooding with magnitude similar to that of a hurricane. They are more common in winter than in summer, based on changes in the steering winds.

What about a polar vortex bomb cyclone? Is that a thing? Yes and no. It is a thing that gets your attention. It is not a new thing, though. It is actually two different things. These are words that meteorologists have used for decades, and in some cases, centuries. All professions have jargon that would make you say, "Huh? What's that?!", and it's often jargon that's better left within the profession. Here are words and phrases you hear in winter:

A cyclone is any low-pressure storm system that spins. "Cyclone" has different meanings around the world. You hear it applied to tropical storms, hurricanes, tornadoes, and winter storms. Even the rollercoaster at Coney Island is called Cyclone, but that's a little different.

A vortex is just what it is called. It's something that spins too. With tornadoes, you hear "multiple vortex." That's when one large tornado has tiny ones spinning around it, or when several small tornadoes from one thunderstorm all spin around a common center, with the effective result of one larger tornado.

"Polar vortex" may sound like the stage name of an exotic dancer, but it is just air that circulates around the North Pole. More correctly, it is a wind at high levels of the atmosphere, like the jet stream, which encircles the planet near the poles. In the northern hemisphere, it acts as the boundary for Arctic air. We don't feel the polar vortex at the ground but as it meanders really far south, we feel the frigid air it allows to enter the United States; and when that air arrives, it's not always windy at ground level. It's what used to just be called an Arctic outbreak.

What about "bomb cyclone"? That's pretty much a winter storm. The "bomb" in front of cyclone is used by a meteorologist for a winter storm that rapidly loses pressure—at least 1 millibar per hour, for 24 hours. In any storm, as pressure falls, winds increase. A regular cyclone or winter storm can be stronger with larger impact, but bomb cyclone just sounds cool, doesn't it?! The "bomb" designation simply means it strengthens very fast. A bomb cyclone is nothing new. Meteorologists started using that phrase in the 1970s. Related to "bomb cyclone" is "bombogenesis," which means the beginning of a bomb cyclone.

What about "nor'easter"? That is a cyclone. It could be a bomb cyclone but doesn't have to be. A nor'easter is a winter storm that moves up the eastern seaboard of the United States. As it moves through New England, the winds blow back toward the coast from the northeast. That's where it gets its name.

Snow squall is simply a burst of snow, which may not cover a large area, but rapidly lowers visibility and accumulates. A snow squall is like the winter equivalent of a summer downpour except that it involves snow, rather than rain.

Clipper is a fast moving little low-pressure system that crosses multiple states and leaves a band of snow that could be several inches to over half a foot. It gets its name from clipper ships which were small, fast sailing ships.

Black ice is not black. It is clear. Black ice is the nickname for a thin coating of ice on roadways that is hard to see. You see the roadway color beneath it, which is often black. Ice is hazardous for motorists, but when they can't see it and are suddenly upon it, it is that much more dangerous. It's a hazard for pedestrians too.

Social media and traditional media are good at latching onto professional jargon when it makes a unique headline. Even if the words are new to you, the science of an atmospheric river, bomb cyclone, polar vortex, nor'easter and blizzard are old.

Atmospheric river is a phrase that's been around since the 1990s. When atmospheric moisture, or water vapor becomes concentrated in a narrow band that extends across oceans or continents, it's known as an atmospheric river. Atmospheric rivers happen around the world. Just like the jet stream is a concentrated band of wind, within the steering winds, an atmospheric river is a ribbon or plume of air with much higher moisture content than surrounding air. It carries freshwater moisture from the tropics to higher latitudes. To a meteorologist, that plume is at least several hundred miles wide, and it could be well over a thousand miles long.

Over the Pacific Ocean, atmospheric rivers are frequent in winter, varying in strength and number, as a normal seasonal cycle. A nickname for an atmospheric river that flows from the Hawaiian Islands to the west coast of the United States is Pineapple Express.

Atmospheric rivers are important because they deliver large amounts of rain and snow to land which can benefit agriculture and water resources, or they may overwhelm communities when they stall, resulting in flooding of rivers, lakes, and streets. On the west coast of the United States, atmospheric rivers also deposit tremendous snowfall in higher elevations. That also may be good or bad, depending on how fast the snow falls and/or melts. A large snowpack acts as a reservoir for fresh water in spring and summer, but too much rain and warm air can melt much of a snowpack within weeks or days. Going from mud to flood, landslides and mudslides may happen in mountainous areas. Heavy precipitation from atmospheric rivers often comes along with a low-pressure storm system, sometimes called a cyclone, which can generate damaging wind equal to a tropical storm or a hurricane. All of that creates challenges for travel.

Because of such a large impact from atmospheric rivers crossing the Pacific Ocean where there is limited weather data, the U.S. Air Force Reserve Hurricane Hunters fly winter missions to drop data buoys into the ocean, and gather more weather and ocean information, so that scientists can better understand and predict how and where atmospheric rivers form, and how long they may last.

Imposter snowflake

What we think of as winter weather can happen in the fall into December. Late December may present a Christmas crisis of dendritic dimensions—a mutation of meteorological magnitude. It is the time of year when some of us are blanketed with a blizzard of Christmas and holiday cards featuring snowflakes. What joy doth flake of snow bring to those starved of a Norman Rockwell idyllic winter scene? Oft times, something is awry that may get past your eye. I'm referring to artistic renderings of snowflakes that are inaccurate, in a science sort of way. I need to draw your attention to imposter snowflakes.

It's a "snowfake" scam that's easy to fall for. It's like being duped by counterfeit thousand-dollar bills. Those are things most of us don't get our hands on. In the warm regions of the U.S., where snow is rare, it's understandably hard to know that snowflakes are typically clumps of ice crystals, while individual snowflakes can be single ice crystals. A single ice crystal has symmetry, but they are not 8-sided as many flakes are depicted in popular culture.

Ice crystals are frozen water, aka H_2O. The basic shape of an ice crystal is a hexagon (6 sides), not an octagon (8 sides). Ice crystals, just like the ones in your freezer, or on your car or window on a frosty morning, can grow and sprout dendrites along their 6 arms, resulting in ornate hexagons. This you might notice if you scrape some freezer frost, put it on a dark cloth and look at it with a magnifying glass, without letting your hot breath melt it. You'll marvel at these crystals the next time you get natural frost. On your vehicle or windowpane or even blades of grass, ice crystals may be individual hexagons, or they may be in a long, branching, and curving line that connects a bunch of hexagons.

Now that you are aware, have some fun and see how often you can spot the counterfeit snowflake of more or less than 6 sides, on Christmas cards and displays. With all fairness to designers and graphic artists, I agree that it is easier to draw a snowflake with 4 or 8 sides, and it is true that Fred Flintstone only has 4 fingers on each hand, and no one seems bothered by that. I'll be ice crystal clear. Yes, Virginia, there is a hexagonal snowflake, and no two are identical. I know, you're saying, "Alan, you have a little too much time on the five fingers of your hands." To that, I offer a frosty stare.

CHAPTER 12

Weather Forecasting

Why wouldn't a meteorologist go to a restaurant on the moon?
There's just no atmosphere.

When you think of weather forecasting you think of meteorologists. A meteorologist is a person trained in the science of meteorology, but a meteorologist doesn't just present the weather on television. In fact, most meteorologists are not on TV. Most meteorologists work for national weather services, armed forces, research labs, forecasting companies, private companies, and for universities as professors. Many meteorologists do not forecast weather!

Some meteorologists specialize in forensic meteorology—that's reconstructing past weather by using satellite, radar, hourly reports and other data. Forensic meteorologists frequently play a role in legal cases where weather may be a factor. That could be anything from a vehicle accident on a wet or foggy road, to a claim of wind damage or flood damage from an intense storm. It may be a facility challenging poor construction of a building where the designer or contractor did not account for local weather extremes. Other meteorologists compile and monitor weather data and create forecasts for estimating energy usage; ideal conditions for agriculture; flight conditions for airlines; sea conditions for shipping companies; snowfall for ski resorts; weather scenarios as they relate to warfare; or for efficient and safe building design in city planning or large construction projects. Meteorologists are found at large outdoor events for public safety from hazardous weather. Meteorologists may do research outside or in laboratories in many locations around the world, or they may create weather models.

A typical meteorologist has a 4-year college degree in meteorology, which is also known as atmospheric science. The degree requires a solid foundation in chemistry, physics, computer science, calculus and statistics before the student starts studying specific meteorology courses. The science of meteorology encompasses all of these disciplines, so the last two years of degree study focus on understanding the physics unique to weather. Upper-level students in meteorology learn how heat is transferred, how to read weather charts, how to forecast weather, computer simulations of weather, how clouds grow and interact, weather instruments, radar and satellites and their limitations, and how light interacts with air and water. Meteorology students may take courses specific to agricultural weather, air pollution, energy resources, winter weather, tropical weather, aviation weather, fire weather, atmospheric electricity, and severe weather.

Some students go on to earn a master's degree, and then a doctorate in meteorology. Other students prepare for careers in broadcast meteorology. They practice and learn how to be expert communicators of atmospheric science to the general public. Broadcast meteorology combines art, performance, and weather.

The Occupational Outlook Handbook from the U.S. Bureau of Labor Statistics is a very good publication to help you plan a career in meteorology.

While meteorology is a fascinating, rewarding profession, know that meteorologists who forecast weather for the public, industry, private companies or even the military may have to deal with mental and physical stress. At work, have you been told to do less? Haha. I know the answer. Let's laugh together at that. Laughter is a good stress-reliever. Have you been given shorter work hours and a raise? Okay, stop laughing. It seems that all of us work harder and juggle more tasks than ever. For a meteorologist, just creating an accurate and precise forecast or assessment on schedule can be stressful.

Big weather events are even more worrisome. It may be long hours and many days of preparation and/or recovery. Storm impact can be costly, tragic, and deadly. If disaster follows, how might you handle it? After an immediate crisis, it wouldn't be unusual to experience sadness, guilt, and depression, to some degree. Combine those with stress from daily life and daily workload, and maybe from feeling overworked and/or underpaid, or from working in an environment that is unhappy, and you might need help. Those stresses, even without a disaster, can grow to critical levels. Stress and depression carry over from home life to job life and vice versa. They may lead to, or be related to, substance abuse. Our jobs certainly influence our overall state of mind. To be the best we can be as weather experts and human beings, we must be aware of our psychological well-being. We all have different tolerances for stress and anxiety, and different methods of handling them. If you are approaching your limit, don't ignore it, address it, no matter what your profession or position.

Imagine the additional stress coming from a barrage of negative comments to you as a broadcast meteorologist who might interrupt a program for a tornado warning, or be criticized in public or on social media for your forecast, your voice, or your appearance.

TV meteorology career

Broadcast meteorology is also called weathercasting. It is a mix of meteorology, computer science, and unscripted presentation. In addition to broadcasting weather information, weathercasters find themselves as scientists, educators, public speakers, celebrities, and participants in charity events. They also frequently perform other duties for their stations such as reporting, editing, photography, or web page or social media maintenance. The profession can be both challenging and rewarding in terms of personal satisfaction and financial compensation. Salaries for experienced weathercasters are good to excellent, based upon common standards. However, starting salaries, for a college graduate, are low, with a typical range of $28,000 to $35,000. Work hours are anything but "9 to 5," and job security can be tenuous. You are sometimes blamed for weather that people don't like; some people become upset or irate when you interrupt their program for a tornado warning; you lose your privacy; you may be called in to work on your time off; and your job may take you far away from friends and family.

There are strict deadlines to adhere to, and the fact that weathercasters serve thousands or even millions of people by communicating information that can make or break a family outing or even save lives, adds to the pressure of the job. For a passionate weather communicator, the easy part of the job of a TV meteorologist is talking and teaching about weather, and loving it. For many people, that is also the hardest part because there is no script. You must deliver a lot of information in a short period. When you balance that with the good you do for people in helping them plan their days, safeguard their families, along with public education and community service, it is a unique profession.

Forecasting

A weather forecast is a prediction of something that has never happened before. All sorts of weather patterns have happened, but not in the precise combination of what you will experience tomorrow. In fact, tomorrow has never happened. When tomorrow does happen, it becomes today! That makes the science of projecting atmospheric details of a day in the future a challenge. Weather forecasts have improved and continue to do so. Reflect on where hurricane forecasts from the National Hurricane Center (NHC) were in 1997, compared to where they were 25 years later. In 1997, the average amount of error in a 3-day location forecast of a tropical storm or a hurricane was around 265 miles. By 2022, the average amount of location error in a 3-day tropical forecast was down to 105 miles. In 25 years, the average 3-day location forecast became 2.5 times more accurate.

A 5-day hurricane location forecast now, is more accurate than a 3-day forecast was 25 years ago. A 5-day hurricane location forecast now, is also as accurate as a 2-day forecast was 25 years ago. That is outstanding progress. It's a tremendous achievement enabled by research, technology, modeling, stronger computers, and experience. That forecast improvement is like taking the field goal record of a kicker at 30 yards, and then finding another kicker who is consistently better at 50 yards, and nearly as good as a kicker at 20 yards. Remember that statistics are averages. All kickers and prognosticators will miss, from time to time.

In 2003, the National Hurricane Center forecast cone for tropical depressions, tropical storms, and hurricanes increased from 3 days to 5 days, thanks to the improvement in forecasting. In 2018, the Hurricane Center started doing experimental 7-day forecasts, but not for public release. I haven't seen the accuracy of the data, but I would expect a 7-day forecast cone to become standard within a decade. This would benefit communities, countries, companies, and agencies, by giving the option of additional pre-planning. The downsides of a 7-day forecast cone would be the confusion generated when multiple cones overlap, and the amount of change that might occur on the 6th and 7th days, along with more time to worry and wonder, procrastinate, and forget that in the 7 days that we are tracking a storm, other storms can form from nothing.

Weather models

Forecasting weather starts with a full set of data describing meteorological and oceanographic parameters over a wide area of the Earth, and up through the troposphere at multiple levels. The simplest forecast takes the current state of the atmosphere and moves weather systems forward at the same rate of speed, direction and growth or decay that they possess. That's known as a persistence forecast because it assumes things will persist on their current trend. This has some use, but it's not the best way to project the future. As all elements interact, physics, chemistry, thermodynamic and hydrodynamic principles are integrated in calculus and differential equations to produce likely outcomes. It is a tremendous number of calculations that do a remarkable job of simulating the workings of meteorology.

Extreme weather events like flooding, tornado outbreaks, and hurricanes are difficult to predict on three levels. One is that we don't really understand what causes all weather extremes. Two is that models may not have enough of a database of extreme events to reasonably portray one. The third difficulty is that sometimes models do project something that has never happened before, and it's just hard to believe by a human forecaster. Getting people to understand the uncertainty is critical, as few weather forecasts have a clear yes/no result. Aside from the probability of something happening, a prediction indirectly includes the confidence level of the forecaster based on weather history and daily data. Rare or extreme weather events always have a lower forecast confidence than routine events.

After extreme weather conditions, you may hear something referred to as a hundred-year storm, or thousand-year storm. Those are often misinterpreted by the public, where people think that the storm can only happen once every hundred years, or once every thousand years. That's not what those phrases mean. If you take a hundred-year storm, for example, it means the statistical odds of that storm happening in ANY year is 1 in 100. Once it happens, the odds reset so that you can get two in one year, or three over several decades. They are not allocated with a limit or preset frequency. If a type of storm were to become more frequent than the odds suggest, the odds would be recalculated, and that data would be incorporated into weather models.

No one weather forecast model is always better than others. Each has biases which make them better on any given day, region, season, and situation, but the problem in forecasting is even if one model does well for a couple of days it doesn't mean it will continue to do so the next day. Models use calculus, statistics, the laws of physics, thermodynamics and hydrodynamics, along with input data to project the future. If you look online, you'll easily find a dozen different ones from various countries, agencies, and universities. If you have a strong enough computer, you can actually run your own model. All model forecasts are only as good as the quality of data that goes into them and our understanding of the atmosphere.

Individual weather forecast models have weaknesses and strengths. Some models are better than others in different latitudes, seasons, months, and scenarios, but no model is on target all of the time. Think of the models as you do stocks. Ask a broker which is the best stock to buy, and she'll tell you which one has done well recently, but you won't get a definitive answer as to which stock will make you rich because that is unknown.

Some stocks do well in different economies and different scenarios of world events. Many brokers will tell you that a blend of stocks is the safest route. As it turns out, a blend of computer models also gives a more reliable and consistent forecast for tropical weather. The ratio of which type of models you use controls the outcome too. That portfolio of prognostication is called an ensemble. Even computer ensembles are not always accurate, but they outperform single computer models, and you can take that to the bank!

Folklore and forecasts

Models act as guidance for a meteorologist to understand the direction of weather change. There are rules of thumb, and folklore, based on observation that may help with short-term forecasts, but they can't be relied on as the only possible scenario. Some folklore doesn't work for forecasting!

"Red sky at night, sailor's delight." That's an old saying that tells us that when the evening sky turns red, the next day's weather will be calm and pleasant. It works often, but not always. It works if you are in a weather pattern where things are moving from west to east. "Red sky at night ..." doesn't work as well in the summer when the steering winds are out of the southeast or when the weather pattern is stagnant. There are many times when a red sky at sunset is just from a break in the clouds to the west, not an entire change of weather. Smoke and ash from wildfires also tint the sky red.

"A ring around the sun or moon portends rain within 48 hours." It may. That ring is a halo. It tells you cirrus clouds are high overhead. Cirrus clouds often precede a warm front in the middle portions of the United States. Now that many locations are under major aircraft corridors, there are more cirrus clouds than ever, as a byproduct of aircraft vapor trails. That old saying doesn't say as much as it used to! Halos are common, anytime.

There's weather folklore about the actions of animals, but animals respond to change rather than predict change. Some creatures are more sensitive than we are to pressure, humidity and even the electric field around a growing thunderstorm, so they may sense something before we do.

Can a groundhog give you clues as to when spring arrives? No! There are many reasons why not, starting

303

with the idea that spring does not have a single definition, along with the fact that seasons always have gradual and subtle starts and ends. Spring, like all seasons, is not an off/on switch. Spring is a transition from winter to summer!

Does a lot of sea life showing up at the shore mean a tropical system will approach? Not as a rule. It's more likely that the creatures found a good source of food or that the water currents simply moved them.

Does a big pecan, almond or walnut crop mean the next season will be wet, dry, or stormy? Take your pick of the nuts, or the supposed outcome. Crops grow based on past and current weather, not future weather.

There are weather misperceptions. When I was a kid, during the NASA Apollo missions, I remember my mother saying, "Whenever the rockets go up, it rains." That was a coincidence. A perception like that is based on people stopping to focus on what is going on around them, during a big event. We remember the weather from big happenings in society and in our lives, like weddings, funerals, and graduations. Just because an unusual or extreme weather pattern led up to or followed something memorable, it doesn't mean that that event caused the weather, or vice versa! People make the mistake of confusing a coincidence with cause and effect. If you had a hot summer and the following winter was very cold, that is not cause and effect. If you get a thunderstorm over your home every Saturday for three weeks in a row, the day did not cause the thunderstorm. That is coincidence.

I'm always amused when people tell me they can forecast weather better than a meteorologist. Some are the same people who think they can outperform a professional athlete. Taking a guess on which way a storm is going, after having watched meteorologists for days, and looking at weather apps is truly a guess. It's not a forecast. A forecast is based on data, using rules of physics, not just a feeling or a hunch. There are many rules and patterns that do make general forecasting for a few hours not too hard, just by using your senses and following trends. However, there are many things that don't work in forecasting, like assuming every weather event will evolve as the previous one, with the same impact.

The outlook on the forecast

In forecasting weather, people always want more information, beyond the week ahead. Online and on TV, you regularly see extended forecasts for a week to two weeks. In any forecast, farther out in time is always fuzzier. Farther is less certain. There's little in life you can be more certain of that is farther away than something that is nearby. That's why meteorologists issue weather outlooks. An outlook is not a specific forecast or a guarantee, but it is what meteorologists use when we see that ingredients for something will be in place. That's the same as for hurricane season when you hear a daily tropical weather outlook, which previews possibilities up to 7 days forward. It's also the same concept used in seasonal outlooks.

Advancements in radar, satellite, and forecasting allow meteorologists to see things or the potential development of things farther out in time and distance. Particularly in spring, you are presented with periodic outlooks for the threat of severe thunderstorms, using the five-level scale of Marginal, Slight, Enhanced, Moderate, and High. Consider that the outlook for severe thunderstorms can go out as much as 8 days, before the actual threat. It didn't always go out that far because forecast ability was limited.

Whenever we are faced with a threatening weather situation, there are two other terms that are analogous to Outlook and Forecast. The first is Watch. Watch is like a short-term outlook. A watch means that ingredients for a certain type of hazard will be in place over a general area in the near future, and you should watch for the threat that could develop.

A Warning is more like a forecast. A Warning says that a specific type of threat is very likely or happening in a specific area, and you've got to act to protect yourself.

Outlooks, Watches and Warnings, all appear as an outline or polygon. All of them can and will change and be adjusted or cancelled, based on data. It is critical to understand that forecasting weather is identical to identifying what a distant vehicle is when you are driving down the interstate. The farther away it is, the less certain you are of what it is. It may not be in your lane. It may be parked. It may exit before you get near it.

Aside from the improving science of detecting and forecasting weather, population growth and expansion means that all weather threats will have increasing impact. Combine that with many more media outlets and social media platforms, and yes, you are hearing more outlooks, forecasts, watches, and warnings than ever before. People confuse outlooks, watches, and warnings, so let's look at this as if a tornado were a tomato.

Tomato warning

There's a Tomato Warning, but only on paper. Don't be alarmed, unless you are allergic to tomatoes. You might be envisioning some streaming movie about tomatoes plummeting from the sky, striking children on a playground, smashing into cars, while painting windshields red to obscure drivers' views, leading to panic and mayhem. This is not that. This is about periodic threats of severe thunderstorms and tornadoes, not tomatoes, often in spring.

Communicating weather hazards is challenging, because even a technical phrase like Tornado Warning will be interpreted differently by the public, depending on what happened or didn't happen the last time there was a Tornado Warning for your area. The same is true when we go under a Watch for severe weather, or when there's an Outlook for severe weather. None of those guarantees a specific outcome. None of those promises the same outcome each time. Let's use the tomato to make this simple.

You go to the garden store, buy some tomato seeds, gather soil, and a planter, and prepare to sow the seeds. You plan for a harvest of buckets full of tomatoes. There's one problem. The tomato seeds in the packet have zero chance of becoming tomatoes without further steps. However, you maintain an Outlook for eventual tomatoes. That's the analogy to an Outlook for tornadoes. An Outlook for tornadoes says that ingredients will become available in sufficient quantity such that if they blend together correctly, there is some future possibility of a tornado. Just as in tornado formation, if ingredients are not mixed to nature's satisfaction, there will be no tomato.

You put the tomato seeds in the soil, add water, leave the planter in the sun, and wait. Within a few days, a seedling emerges. That's not a weed, is it? Nope. Your anticipation builds. Now, we have a Tomato Watch. You are watching and preparing for the possibility of a tomato on and around the tomato plant. There's no certainty that a tomato will emerge. Even when a bud produces that first little green ball and you start to feel proud of yourself, there's a possibility that bugs, critters, weather, and plant disease can prevent a tomato from forming. That's why it's called a Watch. You literally watch, and study how it develops.

A few weeks later, a tomato on the plant! We have a Tomato Warning. A tomato has been spotted. You take a picture for social media, and then prepare to eat the fruit of your labor. The tomato will move toward your mouth at 5 inches per second. Is it bitter, sweet, tangy, or sour? Take action. You won't know the impact until the tomato strikes your tongue. Prepare and have a plan, and a fork!

What's your weather type?

It seems to me that there are two types of people when it comes to dealing with bad or hazardous weather threats: Type A is those who prepare for it, while Type B is those who wait for it to happen and then try to handle it. Naturally, I'm a Type A, but let's look at two people: Esmerelda and Saul.

Esmerelda listens to a weather forecast at least twice a day. She plans her activities to avoid getting caught on slippery roads. When she hears of a Winter Storm Watch,

Esmerelda goes to the grocery store the day before, just in case things turn ugly. During bad weather, she waits until the worst is over before going to run errands. She is Type A.

Saul, on the other hand, doesn't believe weather forecasts. He feels he can do as good a job as any meteorologist because he spends a good deal of time outdoors. When Saul hears of a Winter Storm Watch he shrugs his shoulders and says, "So what? The last time they forecasted a big storm it didn't happen. Even if it does happen, I can handle it." Saul is Type B.

Esmerelda doesn't know the meaning of every single word that the weather presenter on TV uses, but she does know what the difference is between an Advisory, a Watch, and a Warning. Esmerelda, like other Type A people, realizes that if you hear the same words year after year to describe something that has a big impact on people, it is wise to learn what they mean.

Saul will gladly tell you how an Advisory differs from a Watch or a Warning. Ask him. He'll give you an explanation with conviction. Ask him next week. He'll give you a different explanation with just as much conviction. In fact, Saul is so sure of himself that he's willing to go to the library to find meteorology textbooks to prove Esmerelda wrong. It turns out that on this particular winter day there's a Winter Storm Watch.

Esmerelda hears this on the radio and tries to back out of going to the library. Saul taunts her so she agrees to go in the evening with him.

In late afternoon, the Winter Storm Watch is upgraded to a Winter Storm Warning with snow, and high winds developing. Esmerelda backs out of the deal and suggests that Saul not go. Saul does not take her advice. He assumes that the forecasts are going to be, "… wrong as usual." In the 30 minutes it takes Saul to get to the library, an inch of snow falls to add to the ankle-deep accumulation. "This is nothing," Saul thinks to himself, "I don't see why everyone makes a big deal out of it." Saul enters the nearly vacant library. "Great, I have it all to myself!"

Meanwhile, the airport closes, and people lose electricity due to glazed powerlines. Saul is midway through a reference book on weather, when he hears, "Attention library patrons, we will be closing in 15 minutes due to the winter storm. Please bring your books to the checkout …" Saul looks at the clock. The lights flicker twice and then go out. "Uh-oh."

The next morning, after phone service is restored, Esmerelda calls Saul. Saul answers the phone sheepishly, "Ok, you were right, I was wrong—this time." "Saul, I was more concerned about your safety than being right. I'm just glad you made it home." She paused. "So, what did you learn?" "Well, from the book I learned that an

Advisory means weather will be a nuisance to people and make travel a big inconvenience. Advisories are issued for snow, ice and wind chills." Esmerelda injected, "What about a Watch versus a Warning?" "Believe me, I won't get these two mixed up again. The Watch means that a certain type of hazardous weather may occur. It's no promise but it gives you notice to prepare for something that can harm you. The Warning means the bad weather is happening. Take steps to stay out of harm's way." "That's excellent, Saul," Esmerelda exclaimed. "Anything else?" "Yes. From experience, I learned planning ahead is safer." "You know, you just may be my type!"

A precisely accurate forecast

Who wants accuracy in the weather forecast? Who wants precision? Who wants both? Who is confused because you were thinking they were the same?! While they both sound desirable, accuracy and precision are two different things in science. Accuracy is whether something is right or wrong. I'm not alluding to opinions. I'm talking about something that can be objectively measured, like temperature, rainfall, or wind. Precision is the level of detail. I'm not referencing your cousin who tells you every irrelevant detail of a dental procedure they had. Again, I'm talking about a measurement. Who can feel the difference between 50.8 degrees and 51.0 degrees?! Precision is not always required, when we can't sense it.

Something can be accurate while not being precise, and conversely, something can be precise without being accurate. The goal of any weather forecaster is to be accurate and hit the bullseye.

Water freezes at 32 degrees. That number is accurate. If I told you that water freezes at 34.823 degrees, that would be very precise, but also very inaccurate.

When we deal with large numbers, like rainfall for a year, measured in feet and inches, the precision becomes far less important than it is for a single day's worth of rain.

Most folks intuitively get this. It is why gasoline prices are listed to nine tenths of a penny, but the smallest increment you can pay is a penny. Why is it that the nine tenths is always in such a small font? When we drive around pricing gas, we mentally round down, not up. I wonder who benefits from that? I digress.

I see precision fading at the store checkout counter when I pay cash. Many vendors don't make a big deal about pennies. While pennies make the price of something precise, the pennies can get in the way of a swift cash transaction. Pennies also make calculations trickier for some people who didn't have the fine elementary school math teachers that I had. In my occasional fast-food excursion, I can quickly generate a quizzical look from whoever is operating the register when I pay for a $9.37

Number One meal, with a twenty-dollar bill and 2 pennies. I digress again.

Your weather app is very precise, showing hourly projections, and high-resolution computer models, down to the pixel over your neighborhood. Weather apps have more value when you realize that the precision they show in forecasts and graphical displays is not the same as accuracy. That, precisely, is my point.

Accurate forecasts that you see online are often misinterpreted, as there's no person describing exactly what the numbers mean, or exactly when they are for. Many extended forecasts on websites and smartphones are fully automated unless you see them attributed to a person. If you take the rainfall percentage that's listed for a single day, the number alone never tells you the duration, intensity, or total amount of precipitation. It also doesn't tell you if the precipitation is just for a portion of a day vs. the entire day. It is just the chance of any rain falling in one spot nearby. To get the full story and details, you either need to read the full forecast description or listen to a meteorologist on TV.

Even something as simple as a high temperature on a 7-day forecast map can leave questions. Let's say you are looking at a forecast in Denver for 4 days from now, showing a low temperature of 55 and a high temperature of 80. From those two numbers alone, you don't know exactly what time those temperatures occur. You don't know if the temperature rises quickly to 80 degrees by mid-morning and stays there all day, until sunset. It's possible that the temperature stays in the 50s all day, jumps up to 80 for two hours, and then falls back into the 50s.

Weather app forecasts are fully automated, so they will differ from what you hear online or on TV. It gives you a raw computer model value for your exact location, meaning if you drive 5 miles and then look at the app, you'll likely get different values. Within a single county, weather apps generate hundreds of forecast values (something that a human could not do). The values will also update many times per day (sometimes hourly!) so you may see numbers change a lot. While all weather apps look precise, keep in mind that if weather apps were that accurate, and able to communicate how the weather impacts humans, human forecasters would not have jobs! Weather apps are a rough guide to forecast trends. Their greatest value may be in showing radar and alerts for your location.

Never forget that without Wi-Fi or cell service, a weather app won't work, and it certainly can't send you alerts. At times, you may miss an alert or get one for an area that you are in. How does that happen? It depends on whether you are in the alert polygon, and whether the app is set to give you a specific type of alert based on where you set it for. By default, alerting systems send alerts based on your GPS location, only when you are in the alert area. This is to avoid worrying people who are not in the threat area. However, cell phone technology, paired with GPS, is not 100% accurate or 100% reliable, so there will be times where you don't get an alert, even when you are in the polygon. The best thing to do is get a weather radio as a backup.

The coldest day on your weather app

Another shortcoming of some weather apps is highlighted in the heart of winter, when cold fronts and wind shifts shuffle air masses to bring a wide range in morning low temperatures from day to day. It wouldn't be unusual to see morning lows vary by a dozen degrees or more on back-to-back days. Daily low temperature forecasts on some websites and weather apps may confuse people simply because of how the data is displayed. It's not obvious, for a given day, that when a high temperature is stacked above a low temperature, or when a low temperature is shown after a high temperature, which day the low temperature is for.

Some companies that create weather apps use logic that the low temperature for Tuesday, just for example, happens on Tuesday night, so they display that low temperature on Tuesday. They shift the universally accepted 24-hour midnight to midnight day, to a sunrise-to-sunrise period. That's not how most people think. Meteorologically, Tuesday's low happens on Tuesday morning, and that's the way it goes into the historical database. On a calendar, Tuesday night is actually Wednesday. It's like if you were born on July 5th at 1 am, you can't say you were born on the July 4th night!

On an app or website where the low temperatures are not visually separated or staggered between the daily high temperatures, sometimes the only way to know for sure which day the forecast low is for, is to look at the hourly projection of temperatures. Communicating science to non-scientists can be challenging, just as it is challenging for non-scientists to rapidly digest data and graphics and charts.

Weather numbers

Why is it that an anniversary or birthday that ends with a zero is a big deal? We put a lot of emphasis on multi-digit numbers, in multiples of ten, ending with zero. Is that because they are easier to count on ten fingers? There are numbers that are psychologically, legally, or financially significant in our lives. At 10, you become double digits. At 13, you become a teenager. That's a big deal, until you turn 18. At 21, you get all your legal rights. Then, there's a long gap until you reach the age for senior citizen discounts, and then full retirement benefits. Some of these numbers have statistical logic for their importance, but it might seem that numbers and thresholds in weather don't.

Most weather numbers do have an origin that explains why they are odd. Start with freezing and boiling. Water freezes at 32 degrees and boils at 212 degrees, using the Fahrenheit scale, as we do in the United States. In the rest of the world, use of the Celsius scale gives 0 for freezing and 100 for boiling. Those are nice and easy to remember. Both Fahrenheit and Celsius scales do have negative values and many of us like to stay positive. Negative numbers always drag us down. To remain positive, you'd like the Kelvin temperature scale where absolute zero is the coldest that can ever exist in the universe; freezing is 273 degrees; and boiling is 373 degrees. Like the Celsius scale, the difference between freezing and boiling is 100 degrees. There's a good round number again.

Severe thunderstorm wind criterion is 58 mph. That's odd, right? It's not so odd when you realize the threshold is based on the round number of 50 knots, which converts to 58 mph. Maybe it is weird since only pilots and mariners use knots.

If you've ever driven in Canada, on the equivalent of a U.S. interstate highway, you may have gotten excited seeing signs with a speed limit of 100 or even 110. Not so fast, literally! 100 kilometers per hour is 62mph, and 110 kilometers per hour is 68mph.

The starting wind speed of a hurricane is 74 mph. Why not 75 mph? Converting 74 mph gives 64 knots or 119 kilometers per hour. That seems as random as the break points for classes of tornadoes, or categories of hurricanes. All of those numbers are based purely on research results for damage caused by wind. The categories are not as neat as you would want, but they are effective, for ranking.

Rain percentage

In a weather forecast, you'll often hear rain and snow predictions given as a percentage chance. The percent chance of rain is often the statistical odds of measurable precipitation where you are. It does not tell you the intensity, quantity, or duration. It does not tell you if it will only rain once or multiple times over a single day. That means in central Florida, a 40% chance of thunderstorms can result in brief but very heavy rainfall; while in the Pacific Northwest, a 40% chance of rain may be a steady light rain that continues for a few hours. Some predictions use the percentage in a slightly different way—to tell you what portion of a city or county will get measurable precipitation.

That percentage works great for display on a website or weather app, but to avoid an overload of numbers, many meteorologists use words instead. You may hear terms like isolated, widely scattered, scattered and numerous used for showers and thunderstorms. Isolated means 10% coverage; widely scattered is the same as 20% coverage; scattered is 30% to 50%; and numerous is 60% to 70% coverage. If the coverage is very high, then it's called rain (or snow) and described as occasional, intermittent, steady, or widespread. The percentages for precipitation coverage also may apply to the probability of getting wet at a single location, so that scattered showers for an area, meaning 30% to 50% coverage, is just about the same as a 30% to 50% probability at a single spot.

Cloud coverage

You'll hear words like scattered, broken and overcast used by aviators to describe clouds. Each of those equates to a fraction of the sky covered by clouds. These are a matter of science and perspective. For a meteorologist, trying to communicate a picture of what people should expect, the angle of the sun plays a role because a high sun will have more sunlight directly reaching the ground, while a low sun will shine more on the sides of the clouds and/or be blocked by more clouds. To your eye, the same amount of clouds on the horizon, that is over your head, will always appear to be cloudier on the horizon. Whether you spy a field of cumulus clouds dotting a blue sky, your description may depend on whether you are an optimist or a pessimist. It may also depend on your occupation. A roofer wants clouds on a hot day, but an ice cream vendor wants sun.

Final thoughts

Meteorology is weather. It is amazing, incredible, soothing, dangerous, ever-present, fascinating, and in everything we do. Follow patterns closely to learn the cycles of weather that are common. Those who become more observant of our surroundings begin to notice things that were always there. With internet, streaming video, 24-hour cable, satellite TV and social media, we hear about every little weather event that happens, and watch them unfold, live. Most of what happens in weather is not new, it's just that we didn't hear about it in the past or it now affects more people simply because the population has grown or so many more cellphone pictures and videos are shared.

Modern meteorology maintains a mission of understanding the atmosphere and predicting the future. No human being truly knows the future but by studying patterns and processes and the past, we collect clues into future weather. Even when seemingly calm on a cloudy day, weather processes are in perpetual motion. Heat is transferred from sun to ground to sky, and all around. Air molecules move about. Latent heat is stored and released in the hydrologic cycle. Water vapor mixes with air. Condensation floats as clouds. Snow and solid ice sit on cold mountains. Liquid water runs in rivers. We see similar shapes and patterns on different scales in nature, like waves on the ocean and waves in the clouds. The branching of rivers and the branching of lightning. They both gather and carry a substance and are crucial to the health of the planet. Wind forms artistic patterns on snow as ocean waves do the same on sand, of ripples and scallops. That's just as windshear does in air, even though our eyes can't see the patterns in air until clouds make them visible. We feel those patterns as turbulence.

Know that human activity does change climate and it does change weather by mistake. Humanity is inextricably bound to the atmosphere. As you become more appreciative of weather, observe and study the sky at every opportunity and you will constantly find new sights and beauty, fostering a better understanding of how weather impacts daily life, and how your life is intertwined with it.

About the Author

Alan Sealls is an Emmy award-winning broadcast meteorologist. Having worked for decades on television in Milwaukee, Chicago, and in Mobile as a chief meteorologist, you may have watched him years ago on WGN and on CNN. Alan is also an adjunct college professor. He earned meteorology degrees from Cornell University and from Florida State University. He is a past president of the National Weather Association, and he is a Fellow of the American Meteorological Society. In addition to providing weather safety seminars to businesses, Alan is frequently invited to speak at national conferences, and he is often retained by attorneys as a consulting meteorologist or expert witness in legal cases where weather is a factor. Always an energetic and entertaining speaker, Alan has reached tens of thousands of children and adults over his career in public service presentations on the science of meteorology.

Some people recognize Alan for going viral and being proclaimed "Best Weatherman Ever" in 2017 for Hurricane Irma coverage. Alan is also known as an accomplished weather photographer. His weather photos and video work have been used in textbooks, magazines, newspapers, science journals, documentaries, and on network TV programs.

Learn more about weather science from Alan Sealls on YouTube at alansealsweather and on Facebook at alansealsweather and online at alansealls.com

Other books by author

A Career in TV Meteorology

Weather Things in Photos